# Developing Advanced Technologies for Hydrogen Storage and Production

Tolga Taner
*Aksaray University, Turkey*

Hafiz Muhammad Ali
*King Fahd University of Petroleum and Minerals, Saudi Arabia*

IGI Global
Scientific Publishing
Publishing Tomorrow's Research Today

Vice President of Editorial Melissa Wagner
Director of Acquisitions Mikaela Felty
Director of Book Development Jocelynn Hessler
Production Manager Mike Brehm
Cover Design Jose Rosado

Published in the United States of America by
IGI Global Scientific Publishing
701 East Chocolate Avenue
Hershey, PA, 17033, USA
Tel: 717-533-8845
Fax: 717-533-7115
Website: https://www.igi-global.com E-mail: cust@igi-global.com



Library of Congress Cataloging-in-Publication Data

Names: Taner, Tolga, editor. | Ali, Hafiz Muhammad, DATE- editor.
Title: Developing advanced technologies for hydrogen storage and production / edited by Tolga Taner, Hafiz Muhammad Ali.
Description: Hershey, PA : Engineering Science Reference, [2025] | Includes bibliographical references and index. | Summary: "This book covers the energy systems focusing on hydrogen. Focused reviews providing a critical insight into the topic by world renowned experts will be considered. The studies may include, but not limited to the following topic that is determined to contribute to the future in terms of energy and clean environment by developing improved hydrogen production technologies such as electrolysis, thermochemical, electrochemical, photobiological, and photochemical approaches"-- Provided by publisher.
Identifiers: LCCN 2024038215 (print) | LCCN 2024038216 (ebook) | ISBN 9798369308271 (h/c) | ISBN 9798369349205 (s/c) | ISBN 9798369308288 (eISBN)
Subjects: LCSH: Hydrogen as fuel. | Hydrogen--Storage. | Hydrogen--Transportation.
Classification: LCC TP359.H8 D48 2024 (print) | LCC TP359.H8 (ebook) | DDC 665.8/1--dc23/eng/20241125
LC record available at https://lccn.loc.gov/2024038215
LC ebook record available at https://lccn.loc.gov/2024038216

British Cataloguing in Publication Data
A Cataloguing in Publication record for this book is available from the British Library.

# Table of Contents

# Detailed Table of Contents



Fossil fuels were unpopular due to heavy pollution. So demand of sustainable energy is very high all over the world. The Future of Hydrogen offers a thorough and unbiased analysis of hydrogen, outlining current conditions, potential applications, and ways that hydrogen might contribute to the development of a safe, economical, and sustainable energy future A number of technological and economic factors affect the cost of producing hydrogen from natural gas; the two most significant ones are capital expenses and gas pricing. Although it is now widely employed in some industries, hydrogen has not yet reached its full potential in helping to shift to renewable energy sources. It will take bold, focused, and immediate action to break down more obstacles and cut expenses. Hence in this chapter an extensive literature review is conducted to find challenges and priortize most potential challenges by SWARA method.



The transition to a hydrogen-based economy presents both opportunities and hurdles in the pursuit of sustainable energy systems. This chapter explores the multifaceted landscape of hydrogen technology adoption, examining the potential

The cement industry is essential to the growth and development of the Indian economy and infrastructure due to the enormous rise in the housing sector. Any rapidly developing economy faces a lot of energy demand and environmental issues. Green hydrogen is generated through water electrolysis using renewable sources solar, wind, etc playing an important role in resisting climate change. This work describes the implementation of solar energy for the calcination process in cement production. The proposed model have PV power along with battery storage as a key source while diesel generator as back up for emergency requirements. The proposed Chhattisgarh site is located at coordinates of latitude: 21o16.7 N and longitude: 81o52.0 E. This study suggests PV system capacity factor is 18.4% with 4,361 hours/year of operation. The total production of solar energy is 24,234,917 kWh/year. This solar power is meeting the green hydrogen needs of proposed cement plant with 325 kg/year.



The tremendous escalation of the global population, the impending inadequacy of current energy resources, and the alarming levels of environmental pollution have heightened the significance of alternative fuels. This condition requires that the alternative fuels under investigation be environmentally sustainable. This chapter indicates hydrogen, which holds a significant position among alternative fuels due to its availability and environmental sustainability. The enhanced combustion velocity and self-ignition temperature diminish the likelihood of knocking in spark ignition engines. The minimal ignition energy facilitates effortless starting. Due to the absence of carbon in hydrogen, its exhaust emissions are superior to those of fossil fuel engines. The implementation of plasmatron technology resulted in around 25% fuel savings in vehicles, maintaining vehicle efficiency. The accomplishment was realized through the full combustion of fuel in lean combinations. The complete combustion of fuel diminished vehicle emissions and mitigated environmental harm.


*Preeti Parihar, Department of Electrical and Electronics Engineering, SRM Institute of Science and Technology, Modinagar, India*
*Dharmbir Prasad, Department of Electrical and Electronics Engineering, SRM Institute of Science and Technology, Modinagar, India*
*Rudra Pratap Singh, Department of Electrical Engineering, Asansol Engineering College, Asansol, India*
*Ranjeet Singh, Department of Electrical Engineering, Faculty of Technology, Delhi University, India*
*Ariba Rizwan, Department of Electrical Engineering, Asansol Engineering College, Asansol, India*
*Ranadip Roy, Department of Electrical Engineering, Sanaka Educational Trust's Group of Institutions, Durgapur, India*
*Shubham Verma, Software Development Department, Amygb.Ai Pvt. Ltd., Thane, India*

This research, which provides a comprehensive comparative analysis of green hydrogen generation and storage systems, highlights the importance of renewable energy sources in sustainable energy transition. The outcomes show significant variations in economic viability and environmental effect between the intended Cases-1 and -2. The Case-2, with a greater ROI, IRR and simple payback duration, appears to be a more financially feasible option for using green hydrogen. However, it is important to observe the considerable difference in carbon emissions, with Case-2 generating far more carbon than Case-1. The study's proposed locations are Limali, Gujarat and Bandra West, Mumbai, Maharashtra. Applying renewable energy sources such as Wind/PV/GEN, The ROI is 0.4%, IRR is 0.7% and Simple

# Preface

## INTRODUCTION

Hydrogen is employed as a fundamental resource across multiple industries, and it is expected that hydrogen production will significantly contribute to the future development of the hydrogen energy economy. The book "Developing Advanced Technologies for Hydrogen Storage and Production" addresses energy systems centered on hydrogen. Critical reviews offering insights on the topic from recognized global specialists will be considered. The studies may encompass, but are not restricted to, topics aimed at advancing energy and environmental sustainability through the development of enhanced hydrogen production technologies, including electrolysis, thermochemical, electrochemical, photobiological, and photochemical methods. Additionally, membrane separation, pressure swing adsorption, and cryogenic distillation can create hydrogen-rich gases instead of pure hydrogen. This book covers renewable energy, nanotechnology, power generation, boron, military uses, health, food, environmental challenges, and aerospace.

### The Challenges

The initiative for Developing Advanced Technologies for Hydrogen Storage and Production book employs a novel methodology to attain its objectives. This book discusses the advancement of sophisticated technology for hydrogen storage and generation. This document will facilitate the interchange of ideas, best practices, and advances in basic and industrial hydrogen research and engineering. The study of hydrogen and its control covers a wide range of science.

This book examines theoretical and practical engineering science and technology to illuminate hydrogen technology advances. They seek to provide scientists, engineers, technologists, industry experts, and students with relevant and important hydrogen production technology research and regulation data.

The target audience for The Developing Advanced Technologies for Hydrogen Storage and Production will consist of scientists, engineers, technologists, and industry professionals engaged in many hydrogen-related fields.

## ORGANIZATION OF THE HYDROGEN TECHNOLOGY BOOK

The book "Developing Advanced Technologies for Hydrogen Storage and Production" comprises seven chapters, structured into 3 sections to comprehensively address all facets of hydrogen. Each segment comprises multiple chapters that elucidate insights from professional and specialist authors in their respective domains. The sections are given as follows:

## DESCRIPTION OF SECTIONS AND CHAPTERS

### Section 1: Enhancement of Hydrogen Energy

Hydrogen energy systems, in addition to being a green clean energy source and high efficiency, have started to gain importance with new technological developments. Moreover, we can also use hydrogen technology to analyze all discipline and field for the future.

The development of hydrogen energy is the subject of investigation in Section 1, which is broken up into three separate chapters. Challenges and obstacles to the implementation of hydrogen energy are discussed in Chapter 1. Chapter 2 provides an overview of the path that leads to an economy based on hydrogen. It is the advancement of hydrogen energy that is highlighted in this section. Chapter 3 delineates the advancements and challenges in the manufacture of green hydrogen.

### Section 2: Application of Hydrogen Technology in Automotive and Industrial Sectors

The applications of hydrogen technology in the automotive and industrial sectors are discussed in Section 2. In Chapter 4, green hydrogen for cement manufacture is discussed. This chapter places an emphasis on a decarbonization pathway that utilizes hybrid renewable energy sources. Finally, the technology of hydrogen may be quickly implemented in a variety of industries, including the automotive and industrial sectors, as well as other areas. Chapter 5 outlines the hydrogen application of plasmatron technology in fuel preparation and the mitigation of exhaust pollutants.

### Section 3: Renewable Energy Sources and the Green Carbon Cycle: The Future of Hydrogen

In Section 3, the sustainable carbon cycle and renewable energy sources that are relevant to the future of hydrogen are discussed. According to the definition provided in Chapter 6, green computing is a supplemental component in the generation of hydrogen energy inside an integrated system for renewable energy. A discussion on the optimization of integrated power and green hydrogen generation through the utilization of hybrid renewable energy sources is included in Chapter 7. In conclusion, it can be advanced future energy by employing hydrogen technology across all disciplines and fields.

**Tolga Taner**

*Department of Motor Vehicles and Transportation Technologies, Aksaray University, Turkey*

**Hafiz Muhammad Ali**

*Department of Mechanical Engineering, King Fahd University of Petroleum and Minerals, Saudi Arabia*

# Acknowledgment

The editors wish to express gratitude to all individuals participating in this book, particularly the authors and reviewers who participated in the review process. This book would not have materialized without their support.

The editors like to express their gratitude to each author for their contributions. We extend our heartfelt appreciation to the chapter authors for their invaluable time and skills provided to this book.

Secondly, the editor wishes to recognize the significant efforts of the reviewers in enhancing the quality, coherence, and presentation of chapter content. Numerous authors additionally functioned as referees; we much value their dual role.

We thank our appreciation to the entire IGI publishing team for their competent guidance and assistance.

# Introduction

Hydrogen energy systems are considered a green clean energy source with high efficiency. They have started to gain importance with new technological developments. It is believed that they will be the most effective renewable energy system in the future.

Environmental issues like global warming, increased air pollution, deteriorating atmospheric greenhouse effect, and climate change have emerged due to the large amounts of greenhouse gas emissions caused by fossil fuels. Hydrogen is a potential alternative renewable energy fuel source to fossil fuel to reduce greenhouse gas emissions into the atmosphere. Hydrogen is also crucially used as a fuel for generating electricity in fuel cells. Hydrogen is also used as a raw material in some industries, and it is seen that hydrogen production will play an important role in implementing the hydrogen energy economy in the future. It is expected to contribute to the future energy and environmental cleanliness by developing advanced technologies, including methods like electrolysis, thermochemical processes, electrochemical methods, and photosynthesis, to produce hydrogen. In addition, in some cases, hydrogen-rich gases are produced instead of pure hydrogen by processes such as membrane separation, pressure swing adsorption and cryogenic distillation, such as hydrogen separation and purification.

This book presents comprehensive knowledge on the various conventional and state-of-the-art technologies used for all the energy systems to current use in the industry and potential use in domestic applications in the future. The main aim of this book is to highlight advanced technologies for hydrogen production. The book promotes experimental and numerical studies, cutting-edge research, and original work on new technologies and recent developments in this field. Advanced techniques in hydrogen energy, including alternative and renewable energy systems, will be appropriately recognized, with a focus on new technology.

This book covers the energy systems focusing on hydrogen. The book will consider focused reviews by world-renowned experts that offer critical insights into the topic. The studies include the following topics as follows:

- Challenges and barriers in the adoption of hydrogen energy
- Navigating the Path to a Hydrogen Based Economy
- Progress and Obstacles in Green Hydrogen Production
- Green Hydrogen for Cement Production: A Decarbonization Pathway with Hybrid Renewable Energy Sources
- Hydrogen Application of Plasmatron Technology in Fuel Preparation and Mitigation of Exhaust Emissions
- Green Computing for Hydrogen Energy Production as Supplementary Facility in Renewable Energy Integrated System
- Optimization for Unified Production of Electricity and Green Hydrogen with Hybrid Renewable Energy Sources

This book is anticipated to be impactful in numerous academic disciplines and is believed to alter the perception of hydrogen as a future technology. The book offers valuable knowledge that appeals to both academic and professional domains.

Section 1 introduces the Chapter 1 as "Challenges and barriers in the adoption of hydrogen energy", the Chapter 2 as "Navigating the Path to a Hydrogen Based Economy", and the Chapter 3 as "Progress and Obstacles in Green Hydrogen Production".

Section 2 designates the Chapter 4 as "Green Hydrogen for Cement Production: A Decarbonization Pathway with Hybrid Renewable Energy Sources" and the Chapter 5 as "Hydrogen Application of Plasmatron Technology in Fuel Preparation and Mitigation of Exhaust Emissions".

Section 3 highlights the Chapter 6 as "Green Computing for Hydrogen Energy Production as Supplementary Facility in Renewable Energy Integrated System" and the Chapter 7 as "Optimization for Unified Production of Electricity and Green Hydrogen with Hybrid Renewable Energy Sources".

Section 1 was written Chapter 1 by Dr. Suchismita Satapathy, Chapter 2 by Dr. Rahul Uday Urunkar, Dr. Sharad Dattatray Patil, & Sachin Krishnat Pisal, and Chapter 3 by Dr. Sankalp Arpit, & Dr. Tolga Taner.

Section 2 was written Chapter 4 by Dr. Dharmbir Prasad, Dr. Rudra Pratap Singh, Tanuja Tiwary, Dr. Ranadip Roy, & Irfan Khan and Chapter 5 by Dr. Talip Akbiyik, & Dr. Tolga Taner.

Section 3 was written Chapter 6 by Dr. Dharmbir Prasad, Dr. Rudra Pratap Singh, Ariba Rizwan, Dr. Ranadip Roy, & Shubham Verma and Chapter 7 by Dr. Dharmbir Prasad, Dr. Rudra Pratap Singh, Aditya Arya, Dr. Ranadip Roy, & Dr. Sudhangshu Chakraborty.

We would like to thank our heartfelt appreciation to all the authors who have participated in the composition of this book. We are profoundly appreciative of their efforts to enhance the quality and engagement of the book. We anticipate their continued collaboration with us in the future.

We extend our gratitude to IGI Global Press for their patience and understanding throughout the two-year process of developing and authoring this book from inception, as well as for their assistance and support.

# Section 1
# Enhancement of Hydrogen Energy

# Chapter 1
# Challenges and Barriers in the Adoption of Hydrogen Energy

**Suchismita Satapathy**
https://orcid.org/0000-0002-4805-1793
*KIIT University, India*

## ABSTRACT

*Fossil fuels were unpopular due to heavy pollution. So demand of sustainable energy is very high all over the world. The Future of Hydrogen offers a thorough and unbiased analysis of hydrogen, outlining current conditions, potential applications, and ways that hydrogen might contribute to the development of a safe, economical, and sustainable energy future A number of technological and economic factors affect the cost of producing hydrogen from natural gas; the two most significant ones are capital expenses and gas pricing. Although it is now widely employed in some industries, hydrogen has not yet reached its full potential in helping to shift to renewable energy sources. It will take bold, focused, and immediate action to break down more obstacles and cut expenses. Hence in this chapter an extensive literature review is conducted to find challenges and priortize most potential challenges by SWARA method.*

DOI: 10.4018/979-8-3693-0827-1.ch001

## INTRODUCTION

Energy consumption grows across the world with rising population and development. Almost all the globe is dependent on fossil fuels for servicing its energy demands. Although the utility of fossil fuels for satisfying the demand is quite high, fossil fuels are also controversial owing to their destructive influence on the environment. Combustion of fossil fuels (coal, oil, and natural gas) releases massive quantities of carbon dioxide ($CO_2$) and other greenhouse gases into the environment. These gases are the principal source of global warming and climate change. Fossil fuel extraction, processing, and transport can contribute to deforestation, habitat deterioration, soil and water pollution, and ecosystem damage. The burning of fossil fuels emits pollutants such as $SO_2$, $NO_x$, PM, and VOCs. These contaminants are a major source of air pollution and cause substantial health concerns such as respiratory disorders and cardiovascular disease. Air and water pollution generated by fossil fuels have direct and indirect harmful effects on human health, including respiratory illness, cardiovascular disease, cancer, and early mortality. Fossil fuels exploration, mining, and transportation can affect animal habitats, resulting in habitat loss, species relocation, and mortality among wildlife. The use of fossil fuels has long-term repercussions on the ecosystem, such as ocean acidification, increasing sea levels, and altering climate patterns, which create serious concerns for future generations. Despite their historical relevance is a source of power for modern civilization, fossil fuels have numerous substantial downsides and harmful impacts on the environment, human health, and the global economy. Recognizing and tackling the inadequacies of these fossil fuels is vital to shifting to cleaner, more sustainable energy sources and limiting the consequences of climate change and environmental degradation (Mohd Amin & Mohd Zaid 2024). Alternative energy technologies such as renewable energy sources (solar, wind, hydro, geothermal) and energy efficiency measures provide feasible options to lessen our dependency on fossil fuels and build a more sustainable energy future.

By the early 21st century, almost 80% of global energy was derived from fossil fuels. Considering the escalating peril that the Earth's climate presents due to the rising concentration of greenhouse gases in the atmosphere.

Carbon dioxide ($CO_2$) is a principal byproduct of fossil fuel combustion. The use of fossil fuels in industry, transportation, and building has been escalating, greatly augmenting $CO_2$ levels in the Earth's atmosphere. From 1000 CE to the late 18th century, atmospheric $CO_2$ levels fluctuated between 275 and 290 parts per million by volume (ppm) of dry air. By 1959, levels had increased to 316 ppm, and by 2023, they had escalated to 421 ppm. As a greenhouse gas, $CO_2$ captures and re-emits infrared radiation, or net thermal energy, discharged from the Earth's surface (Britannica 2025).

Hydrogen energy refers to the utilization of hydrogen as a diverse energy source and a clean energy carrier. Hydrogen has garnered significant attention as a viable strategy for mitigating greenhouse gas emissions and advancing sustainable energy systems. Hydrogen fuel cell vehicles (FCVs) utilize hydrogen to generate electricity through a chemical reaction with oxygen, yielding only water vapor as a by-product. In contrast to battery electric cars (BEVs), fuel cell vehicles (FCVs) provide emission-free transportation alternatives with extended driving ranges and expedited refilling times (Xu et al. 2025). Hydrogen may be kept for extended durations and subsequently transformed back into electricity when required, offering grid-scale energy storage options to stabilize intermittent renewable energy sources like solar and wind. Hydrogen serves as a fundamental raw material or fuel source across several sectors, including refining, chemical manufacturing, and metallurgy. The shift to low-carbon hydrogen can decrease emissions in several areas (Acar 2024). Hydrogen may be utilized in gas turbines or fuel cells to generate power with minimal or zero emissions, hence facilitating the decarbonization of the energy sector (Khan et al. 2024). Green hydrogen is regarded as the most sustainable energy source, produced by the electrolysis of water and driven by renewable energy sources (Lu et al. 2025).

Hydrogen energy is increasingly favored over fossil fuel energy due to its environmental benefits, energy efficiency, sustainability, and technological maturity. Despite the potential of hydrogen fuel, it remains relatively unfamiliar; thus, issues must be addressed via the formulation of regulations aimed at eliminating fossil fuel pollution entirely and preserving the environment.

## BACKGROUND

A World Health Organization statement (2018) indicates that the utilization of coal, oil, and natural gas contributes to air pollution and climate change, resulting in significant adverse consequences on human health (Tarhan & Çil 2021). Nanda et al. (2021) elucidate that hydrogen energy is a source of power that has garnered significant interest in recent years. It is a renewable energy source that substitutes fossil fuels (Singh et al. 2022). The utilization of fossil fuels, including oil, coal, and natural gas, adversely impacts the environment, contributing to climate change and global warming (Ali et al. 2024). The gravimetric energy density of hydrogen is generally approximately seven times greater than that of fossil fuels (Tarhan & Çil 2021). Yang et al. (2009) assert that hydrogen energy is undeniably a pivotal energy source for the future; nonetheless, certain issues require enhancement. Following the production of hydrogen energy, challenges related to transportation and storage emerge, as noted by Murray (2018). Baykara (2018) asserts that hydrogen energy is environmentally and climatically benign. In the future, hydrogen will serve as an energy source to supplant oil in transportation modes including automobiles, aircraft, trains, and vessels (Farrell et al. 2003). The European Commission predicts this. The High-Level Group for Hydrogen (2003) projected that 35% of vehicles in Europe will operate on hydrogen by 2040. Hydrogen can be generated from a fossil fuel source, specifically natural gas. Hydrogen manufacturing additionally utilizes electrical energy derived from geothermal sources, biomass, wind, or solar energy. Ewan & Allen (2005) indicate that approximately 40-45% of hydrogen production is derived from gas-steam reforming, about one-third from oil reforming, one-fifth from coal reforming, and roughly 4% from water electrolysis (Tarhan & Çil 2021).

## RESEARCH METHODOLOGY

An exhaustive literature analysis is done to identify the obstacles of hydrogen energy, and a questionnaire of 17 questions is prepared and distributed among academic and industrial specialists, who are requested to provide their opinions on a Likert scale of 1 to 3 (1 = disagree, 2 = no opinion, 3 = agree).Subsequently, the obstacles to hydrogen fuel adoption are identified, and the SWARA (Step-wise Weight Assessment Ratio Analysis) method is employed to prioritize these issues (Nithyanandan & Sindhu Vaardini 2023). Table 1 delineates the obstacles associated with the adoption of hydrogen energy.

*Table 1. Adoption challenges of hydrogen energy*

| Sl no | Question | 1 | 2 | 3 |
|---|---|---|---|---|
| 1. | High cost of producing energy | | | |
| 2. | Requires dedicated infrastructure for production, storage, distribution, and dispensing | | | |
| 3. | Requires significant renewable for production | | | |
| 4. | Storage and transportation challenges | | | |
| 5. | Safety challenges in production, storage, and transportation | | | |
| 6. | Technology developments **are needed to improve hydrogen production** | | | |
| 7. | Large-scale energy demands require significant investment and scaling (Özsungur et al. 2025). | | | |
| 8. | Inadequate policy support and regulations may hinder investment and deployment of hydrogen technologies. | | | |
| 9. | Hydrogen must compete with well-established energy sources like fossil fuels | | | |
| 10. | Public awareness and acceptance of hydrogen technologies need to be improved | | | |
| 11. | Clear policies supporting hydrogen infrastructure development, research, and deployment are crucial | | | |
| 12. | Awareness of hydrogen energy benefits is required. | | | |
| 13. | Efficient hydrogen storage and transportation methods | | | |
| 14. | Public perception and regulatory concerns around safety need to be addressed. | | | |
| 15. | Existing methods like compression or liquefaction are energy-intensive and costly. | | | |
| 16. | Building this infrastructure is costly and time-consuming | | | |
| 17. | This cost needs to be significantly reduced to make hydrogen competitive with conventional fuels (Simona et al. 2024). | | | |
| 18. | Pollution free environment | | | |
| 19. | Increasing public understanding and support for sustainable practices. | | | |
| 20. | Engaging stakeholders across sectors to foster circular economic initiatives. | | | |
| 21. | Influence of vested interests in traditional energy sectors | | | |
| 22. | Need for innovation and research in emerging technologies | | | |
| 23. | Cost needs to be significantly reduced to make hydrogen competitive with conventional fuel (Simona et al. 2024). | | | |

## RESULTS AND DISCUSSION

Innovative applications for hydrogen, the provision of clean hydrogen, and infrastructural initiatives are currently at the most precarious stage of the adoption curve. Targeted, time-constrained loans, guarantees, and additional instruments can facilitate private sector investment, knowledge acquisition, and the sharing of risks and opportunities. The SWARA multi-criteria decision-making process is conducted, and the challenges associated with the use of hydrogen energy are prioritized using factor analysis. Table 2 presents the results of the factor analysis. The outcome of the factor analysis indicates that Cronbach's alpha values exceeding 0.5 are identified across ten constructs: cost of production, infrastructure development, energy intensity of production, storage and transportation, safety concerns, technology development, scaling up, policy regulatory framework, competition with established technology, and market acceptance and awareness. The SWARA ranking method is subsequently employed to evaluate the adoption problems of hydrogen.

*Table 2. Factor analysis of the study*

| | Item | F1 | F2 | F3 | F4 | F5 | F6 | F7 | F8 | F9 | F10 |
|---|---|---|---|---|---|---|---|---|---|---|---|
| Cost of production | 1 | 0.512 | | | | | | | | | |
| | 15 | 0.563 | | | | | | | | | |
| | 17 | 0.602 | | | | | | | | | |
| | 23 | 0.616 | | | | | | | | | |
| Infrastructure development | 2 | | 0.516 | | | | | | | | |
| | 16 | | 0.572 | | | | | | | | |
| Energy intensity of production | 6 | | | 0.604 | | | | | | | |
| | 20 | | | 0.552 | | | | | | | |
| Storage and transportation | 4 | | | | 0.523 | | | | | | |
| | 13 | | | | 0.533 | | | | | | |
| Safety concerns | 5 | | | | | 0.526 | | | | | |
| | 14 | | | | | 0.547 | | | | | |
| Technology development | 21 | | | | | | 0.569 | | | | |
| | 22 | | | | | | 0.556 | | | | |

continued on following page

*Table 2. Continued*

| | Item | F1 | F2 | F3 | F4 | F5 | F6 | F7 | F8 | F9 | F10 |
|---|---|---|---|---|---|---|---|---|---|---|---|
| Scaling up | 7 | | | | | | | 0.609 | | | |
| | 13 | | | | | | | 0.632 | | | |
| Policy regulatory framework | 8 | | | | | | | | 0.628 | | |
| | 12 | | | | | | | | 0.619 | | |
| | 19 | | | | | | | | 0.583 | | |
| Competition with established technology | 9 | | | | | | | | | 0.592 | |
| | 11 | | | | | | | | | 0.588 | |
| Market acceptance and awareness | 3 | | | | | | | | | | 0.625 |
| | 10 | | | | | | | | | | 0.681 |

## Steps of SWARA Method

The SWARA method was utilized to establish the weights of the criteria and sub-criteria through the subsequent procedures (Satapathy & Chauhan 2024):

Stage 1: The parameters are arranged in order of relevance, with the most significant positioned first and the least significant last.

Stage 2: The relative relevance of the average value ($s_j$) is assessed in relation to the importance of the parameter $c_j$ compared to the criterion $c_j$+1.

Stage 3: The computation of coefficients ($k_j$) is as follows:

$$k_j = \begin{cases} 1 j = 1, \\ s_j + 1 j > 1 \end{cases}$$

Stage 4: The computations of the adjusted weights ($q_j$) are as follows:

$$Q_j = \begin{cases} 1, j = 1 \\ \frac{q_{j-1}}{k_j} j > 1 \end{cases}$$

Stage 5: The calculations of comparative weights ($w_j$) of the evaluation conditions are as below:

$$W_j = \frac{q_j}{\sum_{k=1}^{n} qk}$$

where, $n$ = number of the parameters.

Table 3 presents the cost of production. Table 4 shows the development of infrastructure. Table 5 demonstrates the energy intensity of production.

*Table 3. Cost of production*

| Sub parameter | Relative significance of average value | Coefficients | Recalculated weights | Last weights |
|---|---|---|---|---|
| 1 | | 1 | 1 | 0.258048823 |
| 15 | 0.12 | 1.12 | 0.892857143 | 0.230400735 |
| 17 | 0.13 | 1.13 | 0.991150442 | 0.255765205 |
| 23 | 0.14 | 1.14 | 0.99122807 | 0.255785237 |
| | | | 3.875235656 | |

*Table 4. Infrastructure development*

| Sub parameter | Relative significance of average value | Coefficients | Recalculated weights | Last weights |
|---|---|---|---|---|
| 2 | | 1 | 1 | 0.555555556 |
| 16 | 0.0.25 | 1.25 | 0.8 | 0.444444444 |

*Table 5. Energy intensity of production*

| Sub parameter | Relative significance of average value | Coefficients | Recalculated weights | Last weights |
|---|---|---|---|---|
| 6 | | 1 | 1 | 0.519230769 |
| 20 | 0.08 | 1.08 | 0.925925926 | 0.480769231 |

Table 6 determines storage and transportation. Table 7 obtains the safety issues. Table 8 emphasizes technical growth. Table 9 demonstrates scaling. Table 10 is the regulatory policy framework. Table 11 is the competition with existing technology. Table 12 shows market acceptance and awareness. Table 13 illustrates the SWARA Rank of Dimensions. Table 14 gives the final rating of the study.

*Table 6. Storage and transportation*

| Sub parameter | Relative significance of average value | Coefficients | Recalculated weights | Last weights |
|---|---|---|---|---|
| 4 | | 1 | 1 | 0.559471366 |
| 13 | 0.27 | 1.27 | 0.787401575 | 0.440528634 |

*Table 7. Safety concerns*

| Sub parameter | Relative significance of average value | Coefficients | Recalculated weights | Last weights |
|---|---|---|---|---|
| 5 | | 1 | 1 | 0.567099567 |
| 14 | 0.31 | 1.31 | 0.763358779 | 0.432900433 |

*Table 8. Technology development*

| Sub parameter | Relative significance of average value | Coefficients | Recalculated weights | Last weights |
|---|---|---|---|---|
| 21 | | 1 | 1 | 0.541284404 |
| 22 | 0.18 | 1.18 | 0.847457627 | 0.458715596 |

*Table 9. Scaling*

| Sub parameter | Relative significance of average value | Coefficients | Recalculated weights | Last weights |
|---|---|---|---|---|
| 7 | | 1 | 1 | 0.545454545 |
| 13 | 0.2 | 1.2 | 0.833333333 | 0.454545455 |

*Table 10. Policy regulatory framework*

| Sub parameter | Relative significance of average value | Coefficients | Recalculated weights | Last weights |
|---|---|---|---|---|
| 8 | | 1 | 1 | 0.36478484 |
| 12 | 0.12 | 1.12 | 0.892857143 | 0.32570075 |
| 19 | 0.32 | 1.32 | 0.848484848 | 0.30951441 |

*Table11. Competition with established technology*

| Sub parameter | Relative significance of average value | Coefficients | Recalculated weights | Last weights |
|---|---|---|---|---|
| 9 | | 1 | 1 | 0.604743083 |
| 11 | 0.53 | 1.53 | 0.653594771 | 0.395256917 |

*Table 12. Market acceptance & awareness*

| Sub parameter | Relative significance of average value | Coefficients | Recalculated weights | Last weights |
|---|---|---|---|---|
| 3 | | 1 | 1 | 0.570815451 |
| 10 | 0.33 | 1.33 | 0.751879699 | 0.429184549 |

*Table 13. SWARA Rank of Dimensions*

| Sub parameter | Relative significance of average value | Coefficients | Recalculated weights | Last weights | Rank |
|---|---|---|---|---|---|
| 1 | | 1 | 1 | 0.166021419 | Ist |
| 2 | 0.58 | 1.28 | 0.78125 | 0.129704233 | 2nd |
| 3 | 0.43 | 1.43 | 0.699300699 | 0.116098894 | 4th |
| 4 | 0.41 | 1.41 | 0.709219858 | 0.117745687 | 3rd |
| 5 | 0.63 | 1.63 | 0.479294479 | 0.079573149 | 8th |
| 6 | 0.21 | 1.21 | 0.577934462 | 0.095949499 | 5th |
| 7 | 0.48 | 1.48 | 0.479202607 | 0.079557897 | 7th |
| 8 | 0.32 | 1.32 | 0.363101878 | 0.060282689 | 10th |
| 9 | 0.21 | 1.21 | 0.477631787 | 0.079297107 | 6th |
| 10 | 0.05 | 1.05 | 0.456383435 | 0.075769425 | 9th |

*Table 14. Final ranking*

| | | |
|---|---|---|
| **1** | **0.258048823** | **19th** |
| **15** | 0.230400735 | **22nd** |
| **17** | 0.255765205 | **21st** |
| **23** | 0.255785237 | **20th** |
| **2** | 0.555555556 | **5th** |

continued on following page

*Table 14. Continued*

| **1** | **0.258048823** | **19th** |
|---|---|---|
| **16** | 0.444444444 | **12th** |
| **6** | 0.519230769 | **7th** |
| **20** | 0.480769231 | **9th** |
| **4** | 0.559471366 | **4th** |
| **13** | 0.440528634 | **13th** |
| **5** | 0.567099567 | **3rd** |
| **14** | 0.432900433 | **14th** |
| **21** | 0.541284404 | **6th** |
| **22** | 0.458715596 | **10th** |
| **7** | 0.545454545 | **6th** |
| **13** | 0.454545455 | **9th** |
| **8** | 0.36478484 | **17th** |
| **12** | 0.32570075 | **18th** |
| **19** | 0.30951441 | **15th** |
| **9** | 0.604743083 | **1st** |
| **11** | 0.395256917 | **16th** |
| **3** | 0.570815451 | **2nd** |
| **10** | 0.429184549 | **13th** |

The SWARA analysis indicates that item number 9 (Hydrogen must compete with well-established energy sources like fossil fuels) ranked first, item number 3 (Requires significant renewable for production) ranked second, and item number 5 (Safety issues in production, storage, and transportation) ranked third. Likewise, among the ten structures, the cost of manufacturing is the most significant barrier that must be mitigated to facilitate hydrogen energy as a clean and widely recognized fuel source. Subsequently, interpretative structural modeling (ISM) is employed to ascertain the relationships among barriers.

## Steps of ISM

### Design Self-Structure Interactive Matrix

where i is the basis to achieve the j factor, the relation is V. Likewise, if j serves as the basis for factor I, the association is A. In both instances, the correlation between the I and j components is denoted as X, whereas the absence of a link is shown by O. Table 15 illustrates the configuration of an Initial Reachability Matrix (SSIM). The reachability matrix is constructed utilizing 1 and 0 for V, X, O, and A (Satapathy 2024). The relation from i to j will be denoted as 1, whereas the relation from j to i will be assigned a value of zero. Relations will be assigned to a value of 1, while the absence of relations will be assigned to a value of 0. Table 17 presents the initial reachability matrix, whereas Table 18 displays the end reachability matrix. Subsequently, the comprehensive driving and reliance value is ascertained. Subsequently, transitivity connections are identified.

### Calculate Level Partition

Following the computation of the reachability matrix, the antecedent matrix is also determined; the intersection set illustrates the common elements of the reachability matrix and the antecedent matrix in Table 15. Subsequent iterative stages are employed to select parameters at each level (Satapathy 2024). Tables 19 through 23 delineate the processes of ISM, whereas Figure 1 illustrates the ISM model pertaining to hurdles to hydrogen energy uptake.

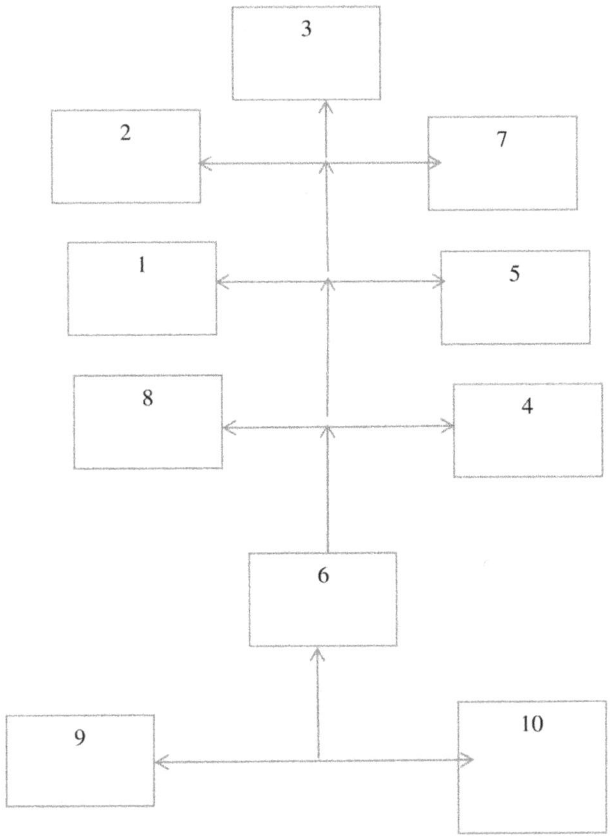


*Table 15. SSIM matrix*

| | C10 | C9 | C8 | C7 | C6 | C5 | C4 | C3 | C2 | C1 |
|---|---|---|---|---|---|---|---|---|---|---|
| C1 | O | O | V | O | O | V | O | A | V | O |
| C2 | O | O | V | A | O | V | V | V | V | |
| C3 | V | V | O | O | V | O | V | V | | |
| C4 | O | O | V | O | A | O | V | | | |
| C5 | O | O | O | O | V | O | | | | |
| C6 | O | O | V | V | O | | | | | |
| C7 | O | O | O | V | | | | | | |

continued on following page

*Table 15. Continued*

| | C10 | C9 | C8 | C7 | C6 | C5 | C4 | C3 | C2 | C1 |
|---|---|---|---|---|---|---|---|---|---|---|
| C8 | V | V | V | | | | | | | |
| C9 | V | A | | | | | | | | |
| C10 | O | | | | | | | | | |

*Table 16. Initial reach-ability matrix*

| | C1 | C2 | C3 | C4 | C5 | C6 | C7 | C8 | C9 | C10 |
|---|---|---|---|---|---|---|---|---|---|---|
| C1 | 1 | 1 | 0 | 1 | 1 | 1 | 1 | 1 | 1 | 1 |
| C2 | 1 | 1 | 0 | 1 | 1 | 1 | 1 | 1 | 1 | 1 |
| C3 | 1 | 1 | 1 | 1 | 1 | 1 | 1 | 1 | 1 | 1 |
| C4 | 1 | 1 | 0 | 1 | 1 | 1 | 1 | 1 | 1 | 1 |
| C5 | 1 | 0 | 0 | 1 | 1 | 1 | 0 | 1 | 0 | 1 |
| C6 | 0 | 1 | 0 | 0 | 1 | 0 | 1 | 1 | 1 | 1 |
| C7 | 1 | 1 | 0 | 1 | 1 | 1 | 1 | 1 | 1 | 1 |
| C8 | 0 | 1 | 0 | 1 | 0 | 1 | 0 | 1 | 1 | 1 |
| C9 | 0 | 1 | 0 | 1 | 0 | 1 | 1 | 0 | 1 | 1 |
| C10 | 0 | 1 | 0 | 0 | 0 | 0 | 0 | 0 | 1 | 1 |

*Table 17. Final reach-ability matrix*

| | C1 | C2 | C3 | C4 | C5 | C6 | C7 | C8 | C9 | C10 | Drive power |
|---|---|---|---|---|---|---|---|---|---|---|---|
| C1 | 1 | 1 | 0 | 1 | 1 | 1 | 1 | 1 | 1 | 1 | 9 |
| C2 | 1 | 1 | 0 | 1 | 1* | 1 | 1 | 1 | 1 | 1* | 9 |
| C3 | 1 | 1 | 1 | 1* | 1 | 1 | 1 | 1 | 1 | 1 | 10 |
| C4 | 1 | 1* | 0 | 1 | 1 | 1 | 1 | 1 | 1 | 1 | 9 |
| C5 | 1 | 0 | 0 | 1 | 1 | 1 | 0 | 1 | 0 | 1 | 6 |
| C6 | 0 | 1 | 0 | 0 | 1 | 0 | 1 | 1 | 1 | 1 | 6 |
| C7 | 1 | 1 | 0 | 1 | 1 | 1 | 1 | 1 | 1* | 1 | 9 |
| C8 | 0 | 1 | 0 | 1 | 0 | 1 | 0 | 1 | 1 | 1 | 6 |
| C9 | 0 | 1 | 0 | 1 | 0 | 1 | 1 | 0 | 1 | 1 | 6 |

continued on following page

*Table 17. Continued*

| | C1 | C2 | C3 | C4 | C5 | C6 | C7 | C8 | C9 | C10 | Drive power |
|---|---|---|---|---|---|---|---|---|---|---|---|
| C10 | 0 | 1 | 0 | 0 | 0 | 0 | 0 | 0 | 1 | 1 | 3 |
| Dependence | 6 | 9 | 1 | 8 | 7 | 8 | 7 | 8 | 9 | 10 | Drive power |

*Table 18. Level partition*

| Challenges | Reach ability set | Antecedent Set | Intersection set | Level |
|---|---|---|---|---|
| C1 | 1,2,4,5,6,7,8,9,10 | 1,2,3,4,5,7 | 1,2,5,7 | |
| C2 | 1,2,4,5,6,7,8,9,10 | 1,2,3,4,6,7,8,9,10 | 1,2,4,6,7,8,9,10 | |
| C3 | 1,2,3,4,5,6,7, 8, 9, 10 | 3 | 3 | |
| C4 | 1,2,4,5,6,7,8,9,10 | 1,2,3,4,5,7,8,9 | 1,2,4,5,7,8,9 | |
| C5 | 1,4,5,6,8,10 | 1,2,3,4,5,6,7 | 1,4,5,6 | |
| C6 | 2,5,7,8,9,10 | 1,2,3,4,5,7,8,9 | 2,5,7,8,9 | |
| C7 | 1,2,4,5,6,7,8,9,10 | 1,2,3,4,6,7,9 | 1,2,4,6,7,9 | |
| C8 | 2,4,6,8,9,10 | 1,2,3,4,5,6,7,8 | 2,4,6,8 | |
| C9 | 2,4,6,7,9,10 | 1,2,3,4,6,7,8,9,10 | 2,4,6,7,9,10 | I |
| C10 | 2,9,10 | 1,2,3,4,5,6,7,8,9,10 | 2,9,10 | I |

*Table 19. Iteration-1*

| Challenges | Reach ability set | Antecedent Set | Intersection set | Level |
|---|---|---|---|---|
| C1 | 1,2,4,5,6,7,8 | 1,2,3,4,5,7 | 1,2,4,5,7 | |
| C2 | 1,2,4,5,6,7,8 | 1,2,3,4,6,7,8 | 1,2,4,6,7,8 | |
| C3 | 1,2,3,4,5,6,7, 8 | 3 | 3 | |
| C4 | 1,2,4,5,6,7,8 | 1,2,3,4,5,7,8 | 1,2,4,5,7,8 | |
| C5 | 1,4,5,6,8 | 1,2,3,4,5,6,7 | 1,4,5,6 | |
| C6 | 2,5,7,8 | 1,2,3,4,5,7,8 | 2,5,7,8 | II |
| C7 | 1,2,4,5,6,7,8,11 | 1,2,3,4,6,7 | 1,2,4,6,7 | |
| C8 | 2,4,6,8,11 | 1,2,3,4,5,6,7,8 | 2,4,6,8 | |

*Table 20. Iteration 2*

| Challenges | Reach ability set | Antecedent Set | Intersection set | Level |
|---|---|---|---|---|
| C1 | 1,2,4,5,7,8 | 1,2,3,4,5,7 | 1,2,4,5,7 | |
| C2 | 1,2,4,5,7,8 | 1,2,3,4,7,8 | 1,2,4,7,8 | |
| C3 | 1,2,3,4,5,7, 8, | 3 | 3 | |
| C4 | 1,2,4,5,7,8 | 1,2,3,4,5,7,8 | 1,2,4,5,7,8 | III |
| C5 | 1,4,5,8 | 1,2,3,4,5,7 | 1,4,5 | |
| C7 | 1,2,4,5,7,8, | 1,2,3,4,7 | 1,2,4,7 | |
| C8 | 2,4,8 | 1,2,3,4,5,6,7,8 | 2,4,8 | III |

*Table 21. Iteration 3*

| Challenges | Reach ability set | Antecedent Set | Intersection set | Level |
|---|---|---|---|---|
| C1 | 1,2,5,7 | 1,2,3,5,7 | 1,2,5,7 | IV |
| C2 | 1,2,5,7 | 1,2,3,7 | 1,2,7 | |
| C3 | 1,2,3,5,7 | 3 | 3 | |
| C5 | 1,5 | 1,2,3,5,7 | 1,5 | IV |
| C7 | 1,2,5,7 | 1,2,3,7 | 1,2,7 | |

*Table 22. Iteration 4*

| Challenges | Reach ability set | Antecedent Set | Intersection set | Level |
|---|---|---|---|---|
| C2 | 2,7 | 2,3,7 | 2,7 | V |
| C3 | 2,3,7 | 3 | 3 | |
| C7 | 2,7 | 2,3,7 | 2,7 | V |

*Table 23. Iteration 5*

| Challenges | Reach ability set | Antecedent Set | Intersection set | Level |
|---|---|---|---|---|
| C3 | 3 | 3 | 3 | VI |

## CONCLUSION

Hydrogen represents the future energy source, providing clean energy with minimal carbon emissions. The current utilization of hydrogen is constrained by inadequate infrastructure and high costs. The cost will decrease as hydrogen consumption increases, and production-storage systems are developed. It is an essential energy source for hydrogen synthesis and storage. Given that fossil fuels utilized for hydrogen production generate carbon dioxide emissions, this method should be discontinued, and hydrogen should be generated using electrical energy derived from renewable sources. In the coming years, generating hydrogen from renewable energy sources (solar, wind, hydropower) will be more economically advantageous. Consequently, hydrogen emerges as a sustainable and clean energy source. The application of hydrogen in the transportation and electrical sectors illustrates its significance as a vital energy source. Hydrogen may be extensively employed as fuel in transportation vehicles in the future. As petroleum usage in automobiles rises, the need for oil will diminish while the demand for hydrogen will escalate. These measurements illustrate the magnitude of the hydrogen economy. Hydrogen ($H_2$) does not occur in its elemental form and must be extracted from compounds such as water or methane. Consequently, it is not an energy source and must be generated using energy. Approximately fifty percent of the annual pure hydrogen production is employed in the synthesis of nitrogen fertilizer via the Haber process, while around twenty-five percent is utilized for the conversion of low-quality crude oil, predominantly sourced from oil sands, which is already a significant chemical commodity. Liquid transport fuel. While it is not as straightforward as natural gas, it can be obtained by working with hydrogen on a large scale. At present, most hydrogen is produced by the steam reforming of either natural gas or coal gas, both of which emit carbon dioxide ($CO_2$). Future demand will focus on hydrogen devoid of carbon dioxide emissions. Strategies to augment hydrogen generation predominantly rely on electrolysis powered by electricity from variable renewable energy sources. Of-peak capacity from conventional nuclear reactors and alternative power sources can also be utilized. A notable prospect for future $CO_2$-free hydrogen production is the direct utilization of heat from nuclear energy to dissociate water using thermochemical reactions facilitated by high-temperature nuclear reactors. The ever-increasing demand for hydrogen from refineries and chemical plants fosters cost-effective technologies. Hydrogen pipeline networks are still limited, resulting in production facilities potentially being located far from consumers. Renewable energy sources, like solar and wind energy, demonstrate intermittent properties. The reliance on solar and wind energy for power generation leads to variability, causing fluctuations and reductions in electricity supply to the system. The remedy for fluctuations in renewable energy generation is the implementation of large-scale energy storage systems. Contemporary generators utilize large batteries to reserve surplus renewable energy for future utilization. Nonetheless, this is only cost-effective if the energy is retained for a finite duration. Over time, the expense of batteries escalates rapidly, rendering the entire process more costly and augmenting the cost per unit of electricity generated. Hydrogen may be relevant in this context. Therefore, the obstacles to hydrogen energy must be reduced.

## REFERENCES


Acar, C. (2024). *Hydrogen storage materials*. Reference Module in Earth Systems and Environmental Sciences., DOI: 10.1016/B978-0-44-313219-3.00113-1

Ali, A., Khan, M. A., & Choi, H. (2024). Supervised Machine Learning-Based Prediction of Hydrogen Storage Classes Utilizing Dibenzyltoluene as an Organic Carrier. *Molecules (Basel, Switzerland)*, *29*(6), 1280. DOI: 10.3390/molecules29061280 PMID: 38542916

Baykara, S. Z. (2018). Hydrogen: A brief overview on its sources, production and environmental impact. *International Journal of Hydrogen Energy*, *43*(23), 10605–10614. DOI: 10.1016/j.ijhydene.2018.02.022

Britannica, Access date: 2025. https://www.britannica.com/science/hydrocarbon

Endo, N., Goshome, K., Tetsuhiko, M., Segawa, Y., Shimoda, E., & Nozu, T. (2021). Thermal management and power saving operations for improved energy efficiency within a renewable hydrogen energy system utilizing metal hydride hydrogen storage. *International Journal of Hydrogen Energy*, *46*(1), 262–271. DOI: 10.1016/j.ijhydene.2020.10.002

Energy, D. European Commission Directorate-General for Research Communication European Commission. High Level Group for Hydrogen, Fuel Cell Technologies, European Commission. Directorate-General for Energy, & European Commission. Directorate General for Research. (2003). Hydrogen Energy and Fuel Cells: A Vision of Our Future: [final Report of the High-Level Group] (Vol. 20719). European Commission, Directorate-General for Research. Unit B-1049 Brussels Fax (32-2) 29-58220.

Ewan, B. C. R., & Allen, R. W. K. (2005). A figure of merit assessment of the routes to hydrogen. *International Journal of Hydrogen Energy*, *30*(8), 809–819. DOI: 10.1016/j.ijhydene.2005.02.003

Farrell, A. E., Keith, D. W., & Corbett, J. J. (2003). A strategy for introducing hydrogen into transportation. *Energy Policy*, *31*(13), 1357–1367. DOI: 10.1016/S0301-4215(02)00195-7

Khan, D. A., Choudhary, A. K., & Sharma, D. (2024). *Hydrogen Energy*. CRC Press. DOI: 10.1201/9781003537816

Lu, H., Xi, D., & Cheng, Y. F. (2025). Hydrogen production in integration with CCUS: A realistic strategy towards net zero. *Energy*, *315*, 134398. DOI: 10.1016/j.energy.2025.134398

Mohd Amin, N. A. A., & Mohd Zaid, H. F. (2024). A review of hydrogen production using $TIO_2$-based photocatalyst in tandem solar cell. *International Journal of Hydrogen Energy*, *77*, 166–183. DOI: 10.1016/j.ijhydene.2024.05.407

Murray, L. J., Dincă, M., & Long, J. R. (2009). Hydrogen storage in metal–organic frameworks. *Chemical Society Reviews*, *38*(5), 1294–1314. DOI: 10.1039/b802256a PMID: 19384439

Nithyanandan, S., & Sindhu Vaardini, U. (2023). A Review on Implementing Lean Construction Technique in Construction Industry. *International Journal of Advanced Research in Science. Tongxin Jishu*, •••, 791–796.

Özsungur, F., Mortaza, C. S., & Hülya, K. B. (2025). *Geopolitical Landscapes of Renewable Energy and Urban Growth*. IGI Global. DOI: 10.4018/979-8-3693-8814-3

Satapathy, S. (2024). Sustainable Supply Chain Management Challenges in Petroleum Transportation. *Advances in Logistics, Operations, and Management Science*, •••, 182–205. DOI: 10.4018/979-8-3693-0669-7.ch009

Satapathy, S., & Chauhan, H. (2024). Sustainable Supply Chain Management in the Mining Industry. DOI: 10.1201/9781003488347

Simona, R. M., Saheb, D., Omar, M., Candin, M. T., & Mustapha, K. (2024). *Advancements in Renewable Energy and Green Hydrogen*. IGI Global.

Singh, R., Altaee, A., & Gautam, S. (2020). Nanomaterials in the advancement of hydrogen energy storage. *Heliyon*, *6*(7), e04487. DOI: 10.1016/j.heliyon.2020.e04487 PMID: 32743097

Tarhan, C., & Çil, M. A. (2021). A study on hydrogen, the clean energy of the future: Hydrogen storage methods. *Journal of Energy Storage*, *40*, 102676. DOI: 10.1016/j.est.2021.102676

World Health Organization. (2018). COP24 special report: health and climate change. Access date: 2020. https://www.energy.gov/eere/articles/hydrogen-clean-flexible-energy-carrier

Xu, Z., Shen, Y., Chen, L., & Wang, C. (2025). High $SO_2$ tolerant and long term reliability of MnSO4-coated Ni-based catalysts for $CO_2$ methanation in coal fired flue gas. *Chemical Engineering Journal*, *507*, 160820. DOI: 10.1016/j.cej.2025.160820

Yang, J., Sudik, A., Wolverton, C., & Siegel, D. J. (2010). High capacity hydrogen storage materials: Attributes for automotive applications and techniques for materials discovery. *Chemical Society Reviews*, *39*(2), 656–675. DOI: 10.1039/B802882F PMID: 20111786

# Chapter 2
# Navigating the Path to a Hydrogen Based Economy

**Rahul Uday Urunkar**
https://orcid.org/0000-0001-8377-9475
*Sanjeevan Engineering and Technology Institute, Panhala, India*

**Sharad Dattatray Patil**
*Rajarambapu Institute of Technology Rajaramnagar, Shivaji University, Kolhapur, India*

**Sachin Krishnat Pisal**
https://orcid.org/0009-0005-9688-5185
*Sanjeevan Engineering and Technology Institute, Panhala, India*

## ABSTRACT

*The transition to a hydrogen-based economy presents both opportunities and hurdles in the pursuit of sustainable energy systems. This chapter explores the multifaceted landscape of hydrogen technology adoption, examining the potential benefits it offers while addressing the challenges that must be overcome. Through an analysis of policy frameworks, international cooperation efforts, and technological advancements, the chapter navigates the path toward realizing a hydrogen-based economy. From discussions on renewable energy integration to the development of*

DOI: 10.4018/979-8-3693-0827-1.ch002

*supportive policy frameworks and global collaboration initiatives, the chapter provides insights into the opportunities and hurdles encountered along the journey towards a hydrogen-powered future.*

## 1. INTRODUCTION

Today's energy and transportation systems heavily depend on fossil fuels, which are unsustainable. The global demand for energy is projected to increase by up to 50% by 2040, according to the International Energy Agency (IEA) (International Energy Agency, 2023). Despite this growth, there's a global agreement that greenhouse gas (GHG) emissions need to decrease to mitigate climate change effects. Therefore, energy security and climate change mitigation are critical challenges. Balancing emission reduction with economic sustainability is essential for the future. The burning of fossil fuels in economic activities drives climate change by emitting greenhouse gases (GHGs) (Lu, 2017). These gases, by trapping heat, lead to respiratory diseases, smog, and air pollution (Environmental Protection Agency, 2023). The energy sector is a major GHG emitter (Lamb et al., 2021). Scientists emphasize the urgency of moving away from fossil fuel consumption for human survival (Urunkar & Patil, 2021a). Transitioning to alternative energy sources is inevitable. For countries like the USA, major carbon emitters, hydrogen and renewables may offer low-carbon energy alternatives in the future (Bridgeland et al., 2022). The global energy sector is transitioning from fossil fuels to zero-carbon sources in the coming decades (Jawerth, 2022). Hydrogen is increasingly vital for decarbonizing various industries, serving as a chemical feedstock and energy carrier (Mcdowall, 2014; Urunkar & Patil, 2021b). It's especially crucial for reducing emissions in sectors like shipping, chemicals, and aviation. Ongoing projects focus on different aspects of the hydrogen supply chain, from production to utilization (Yue et al., 2021). Many countries have tailored plans for a hydrogen economy, offering opportunities for innovation and international collaboration. This chapter reviews the hydrogen economy literature, discussing its potential and challenges for sustainable development, which can inform technology and policy decisions.

## 2. A BRIEF HYDROGEN HISTORY

The concept of hydrogen as an energy carrier and the visualization of hydrogen economy are not novel ideas. Hydrogen is a naturally occurring element and has been used as a crucial input in industries such as ammonia synthesis since around 1920. Before the 1960s, numerous nations employed hydrogen, mixed with carbon monoxide and methane, known as town gas, for street lighting and household energy purposes such as cooking, heating, and illumination. However, serious discussions about the progress of a hydrogen economy began in the 1960's (Bockris, 2013). Following the oil crises of the 1970s, the idea of a hydrogen-centered energy system gained traction, notably with General Motors introducing the term "hydrogen economy" in 1970, especially concerning future fuel sources for transportation. Hydrogen boasts various applications, but its role in transportation stands out as crucial for its wider acceptance, offering effective solutions for reducing emissions and ensuring energy security, a feature not found in most alternative fuels apart from electricity.

In the beginning, hydrogen was seen as a highly effective medium for transporting energy, offering a more economical option compared to transmitting electricity via copper wires. However, with increasing worries about significant air pollution and $CO_2$ emissions resulting from fuel combustion, hydrogen has gained prominence as a superior clean fuel alternative, generating only water when burned. Nevertheless, the narrative surrounding hydrogen as a fuel for transportation, similar to other alternative fuels, has experienced fluctuations over time. During the late 1990s, there was a notable surge in excitement surrounding hydrogen, driven by advancements in fuel cell technology. This enthusiasm led the automotive industry to make overly optimistic projections about the widespread adoption of Fuel Cell Electric Vehicles (FCEVs), fostering high anticipations amongst governments, industry stakeholders, and the general public. However, these anticipations were incomplete during the 2000s. Instead, the discussion surrounding hydrogen was initially overshadowed by a strong emphasis on biofuels. Later, by the end of the 2000s, the resurgence of battery electric vehicles further diverted attention away from hydrogen. The increasing visibility of electric vehicles threatened hydrogen's perceived role as the fuel of the future. However, despite these challenges, ongoing advancements in automotive fuel cell technology, accompanied by reductions in costs, along with increased collaborative efforts between the public and private sectors to develop hydrogen refueling infrastructures, are driving a renewed wave of materialistic tasks for hydrogen FCEVs.

Currently, hydrogen is being recognized as a deliberately essential component in the energy strategies of many countries worldwide. Hydrogen is perceived as an optimal means of energy storage that can be seamlessly combined with renewable technologies, with the goal of supplanting combustion of fossil fuel and fostering sustainable development within the energy segment. Despite the rapid development of hydrogen economy technologies since the 1990's, the widespread adoption of hydrogen technologies still faces several challenges that will require several decades to overcome, particularly the significant but essential initial investment in infrastructure.

## 3. IMPLICATION OF A HYDROGEN-BASED ECONOMY ON SUSTAINABLE PROGRESS

The hydrogen economy holds significant promise for sustainable development, offering a viable alternative to traditional fossil fuels (Urunkar & Patil, 2023b). While policymakers worldwide are exploring wind technology as well as solar technology to decarbonize the energy segment, these sources face challenges due to fluctuations in availability. Hydrogen produced from renewable sources addresses these challenges by offering superior conversion efficiency, high energy density, and storage capabilities, making it a clean and versatile fuel option (Ren et al., 2020). The growing interest in affordable clean hydrogen suggests its potential to serve as a foundation for an alternative, non-fossil-based fuel, contributing to a more sustainable energy future (Noussan et al., 2020).

Hydrogen, a fundamental element in the universe and a key constituent of water ($H_2O$), holds a vital role in the advancement of human civilization. As industrialization progresses, the production of artificial hydrogen has increased significantly, serving essential functions in various manufacturing methods like ammonia and methanol synthesis, as well as in metal production. Moreover, it serves as an energy source for fuel cells and rockets. Undoubtedly, hydrogen is a crucial intermediate product in industrial processes, considering both its production and consumption. However, its significance to sustainable development lies in its inherent properties, notably its large energy density as well as environmentally friendly combustion characteristics. As a clean fuel option, hydrogen has the potential to be widely adopted in Fuel Cell Vehicles (FCV's) and co-generation systems, particularly in industrial and domestic sectors where traditional fossil fuel combustion leads to air pollution. Presently, hydrogen used for energy supply is primarily derived from fossil fuels or surplus renewable energy generation. Despite not naturally occurring as an energy resource on Earth, hydrogen holds considerable promise as an environmentally friendly and efficient energy carrier for promoting the production of renewable and eco-friendly energy, thus contributing to sustainable development.

In the early 1970s, Gregory et al. (Gregory et al., 1972) highlighted the potential benefits of employing hydrogen as a medium for long distance transmission of energy. Compared to transmitting electricity via high voltage lines, transporting hydrogen through pipelines was deemed significantly cheaper and more efficient, excluding the construction costs of infrastructure. However, at that time, the prospect of constructing a dedicated pipeline network for hydrogen transport seemed daunting due to the substantial initial investment required. Recently, advancements have provided more transportation options for hydrogen, including compressed tanks, fuel cells, or repurposing existing pipelines (though this presents challenges such as hydrogen embrittlement). These alternatives support the establishment of affordable regional hydrogen networks (Ball et al., 2015). Moreover, the proliferation of Fuel Cell Vehicles (FCV's) is expected to naturally create a network of hydrogen via road transport infrastructure. There is an expectation that the average cost of transporting hydrogen will significantly decrease in the near future (International Energy Agency, 2023). When considering cost-effectiveness and convenience in energy conversion, hydrogen emerges as a viable option for short and long term energy storage, particularly in situations necessitating large storage capacities.

The development and implementation of hydrogen technologies are widely recognized as crucial for sustainable development. Beyond bolstering energy security and advancing energy efficiency and environmental sustainability, these technologies are anticipated to encourage a new wave of economic expansion and job creation through technological invention and widespread adoption within society, often referred to as a "hydrogen economy." Acknowledging this significance, there has been a surge in interest in hydrogen since the 1990s, with major countries worldwide earnestly developing, deploying, and assessing the feasibility of hydrogen technology. Presently, many significant nations have released roadmaps for the hydrogen economy, guiding research priorities and facilitating the rational allocation of social and financial resources based on their unique circumstances (Mcdowall, 2012). Notably, leaders in research and development, such as Japan, Korea, the EU, and the USA, are spearheading efforts, while countries like China and India are following suit, spurred by their sizable domestic markets.

## 4. HYDROGEN APPLICATIONS, DEMAND AND SUPPLY

### 4.1 Applications

Hydrogen finds diverse applications, including but not limited to powering heavy weight transportation like shipping, electric vehicles fuel cells, production of steel and cement, producing green ammonia to prepare fertilizer, as well as generating power while stabilizing the electricity grid, as illustrated in Figure 1.

*Figure 1. Uses of hydrogen as clean energy*

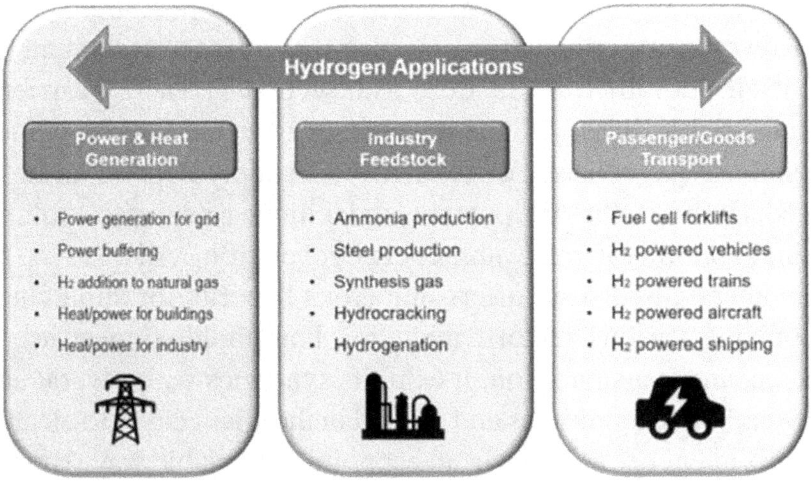


Currently, hydrogen serves as a crucial feedstock in both the chemical and petroleum industries. Approximately 50% of total hydrogen usage is attributed to the production of ammonia for fertilizer manufacturing, followed closely by its role in refining crude oil, which accounts for nearly 40% of its usage. The worldwide hydrogen industrial gas sector is substantial, with total production reaching about 700 billion N-$m^3$ (approx. 60,000 kt), equivalent to fueling over 600 million FCEV's. However, this production is predominantly reliant on fossil fuels, with approximately 50% derived from natural gas and nearly 33% from crude oil fractions processed within refineries. Water electrolysis, accounting for only about 4% of production, represents a minor portion. Emerging applications for hydrogen primarily include its use as a fuel for FCEV's and as a prospective storage solution for electricity generated from random renewable sources like wind energy. Additionally, there is a niche marketplace for movable fuel cells that utilize hydrogen to supply electricity for electronic devices or situations where access to the grid is limited. Looking ahead, hydrogen holds potential for further applications, including heat as well as electricity generation in stationary type fuel cells for domestic or business use, suggesting a broader role in the future energy landscape.

Hydrogen is frequently described as an "energy carrier" or an "energy vector" due to its versatility across a wide range of potential application areas. However, it's crucial to emphasize that hydrogen is not an energy source itself but rather, it functions as a secondary energy carrier, similar to electricity. However, the advantages of using hydrogen in terms of greenhouse gas releases or reliability of supply depend on how it is produced. Hydrogen has the potential to facilitate an energy revolution by providing flexibility in renewable energy generation systems. As a carrier of clean energy, hydrogen offers numerous benefits for simultaneously decarbonizing various sectors, including households, industries, trade, commerce, and transportation. It exhibits synergies with several alternative low-carbon technologies and can aid in the changeover to cleaner and decarbonized energy systems in an economically viable way (Brandon & Kurban, 2017). Furthermore, hydrogen is believed to play a pivotal role in the widespread deployment of eco-friendly energy produced by renewable sources (Noussan et al., 2020).

## 4.2 Hydrogen Demand

The momentum behind the changeover to a hydrogen economy is rapidly growing, supported by a plethora of records from reports and research findings highlighting hydrogen's prospective across various segments of energy systems, including electricity generation (Rosen & Koohi-fayegh, 2016). Assessments conducted by OECD governments further underscore the economic as well as technical feasibility of hydrogen as a promising option for decarbonizing transportation by the use of fuel cells (Hart et al., 2016). In 2021, worldwide demand for hydrogen reached around 94 million metric tons annually. About 50% of this demand was allocated to the production of methanol and ammonia, with the remaining half being used in the manufacturing of iron and steel and petrochemical refineries (International Energy Agency, 2023). Hydrogen also serves as an energy carrier, capable of storing and delivering a substantial volume of energy. Its application in fuel cells enables its use for generation of power, heat, and electricity, with transport evolving as significant possible marketplaces for hydrogen (Satyapal, 2017). Recent inclinations indicate a surge in demand for clean hydrogen, with projections suggesting that demand could possibly twice to in excess of 180 Mt annually by 2030 (Hart et al., 2016). This upward trajectory underscores the increasing recognition of hydrogen's pivotal role in the global energy transition towards a more sustainable and low-carbon future.

## 4.3 Hydrogen Supply

Hydrogen supply encompasses a spectrum of colors, despite its natural colorless state. These colors are assigned based on the production process, resulting in hydrogen being classified as black, brown, grey, green or blue and rarely displaying hues of pink, turquoise, or white as outlined in Table 1.

*Table 1. Categorization of hydrogen based on utilized production techniques (Kochanek, 2022)*

| Type of Hydrogen | Source | Production Techniques | Comments |
|---|---|---|---|
| Brown | Lignite coal | Gasification | The most environmentally harmful option with a significant greenhouse gas footprint. Carbon dioxide and carbon monoxide produced during the process are not captured for reuse or storage. |
| Black | Bituminous coal | | |
| Grey | Natural gas | Reforming of natural gas | Average greenhouse gas footprint. |
| Turquoise | Methane | Pyrolysis | Solid derivative |
| Blue | Coal, Methane | Gasification or Natural gas reforming | Little greenhouse gas footprint |
| Green | Renewable sources such as solar, wind, hydroelectricity | Electrolysis | Negligible greenhouse gas footprint. |
| Purple/ Pink | Nuclear energy | | Negligible greenhouse gas footprint. |
| White | Exists naturally in subsurface deposits and is extracted through drilling processes. | | Negligible greenhouse gas footprint. |

As of today, grey hydrogen dominates the global hydrogen production landscape, resulting in substantial carbon releases. In the year 2021, the making of approximately 94 million metric tons of hydrogen was projected to generate around 900 million metric tons of $CO_2$ (International Energy Agency, 2023) nearly thrice the yearly $CO_2$ releases of France. Also in the same year, Methane reforming, the primary production method, counted for approximately 62% of total production. Next 18% was derived from naphtha reforming in refineries as a byproduct, subsequently utilized for various refinery methods like desulfurization and hydrocracking. Approximately 19% of hydrogen production stemmed from coal gasification, predominantly in China. Only a fraction of hydrogen production, less than 1 million metric ton (Mt) (0.7%), utilized low-carbon pathways, with a mere 35 thousand metric tons (kt) obtained by water electrolysis. Though the volume of hydrogen obtained through electrolysis remained comparatively small, there was a notable 20% increase compared to 2020, indicating a rapid acceleration in the deployment of electrolyzers (International Energy Agency, 2023).

## 5. GLOBAL MARKET OF HYDROGEN ECONOMY

The global market for the hydrogen economy is steadily evolving, driven by the increasing adoption of static type fuel cells for domestic power systems and the advancement of Fuel Cell Vehicle technologies alongside the development of essential hydrogen substructure.

In the coming years, significant market growth is anticipated, spurred by the development of hydrogen based power production and the widespread implementation of hydrogen technology on a large scale. As per a survey report released by Nikkei BP Clean Tech Institute (Nikkei, 2013), the global market size of the hydrogen economy stood at around 80 billion USD in the year 2015. However, projections suggest substantial growth from 100 billion USD in the year 2020 to 1600 billion USD by 2050, as illustrated in Figure 2.

*Figure 2. Hydrogen economy infrastructure market segmented by technology categories (Nikkei, 2013)*

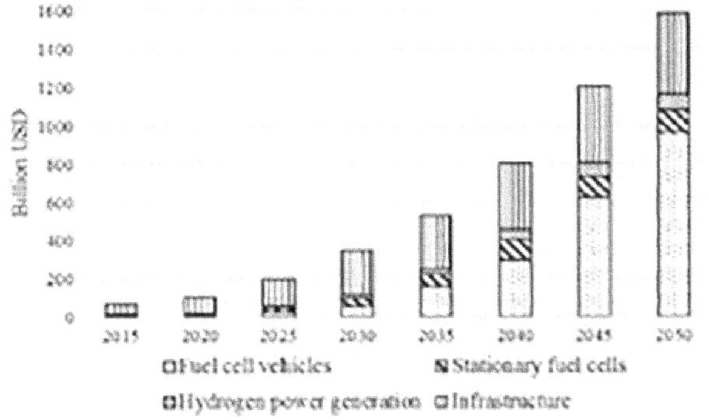


Analyzing the distribution of market scale across different technology categories, it's evident that the construction of infrastructures for applying hydrogen technologies will serve as a significant driver, leading to a rapid increase in market scale to over 300 billion USD by 2030, remaining stable until 2050. This infrastructure development acts as the primary force propelling the hydrogen market forward. Furthermore, with the widespread acceptance of Fuel Cell Vehicles (FCV's) and the establishment of comprehensive infrastructure amenities since the year 2020s, the market size of FCV's is expected to experience a dramatic surge. By exceeding the infrastructure proportion by 2040, the market scale of FCV's is projected to approach 1000 billion USD by 2050, indicating their emergence as the predominant technology supporting the hydrogen economy. Conversely, the market size of stationary fuel cells as well as power generation utilizing hydrogen is predicted to increase steadily, reaching 200 billion USD by 2050.

The distribution of market scale by countries underscores the pivotal role of Europe and North America in driving the hydrogen economy forward. This prominence is attributed to the substantial investments in hydrogen infrastructure, particularly in Europe, where there's a rapid adoption of renewable energy aimed at achieving ambitious $CO_2$ release decrease targets. Conversely, developing nations like China and India are actively engaging in research and development of hydrogen based technologies, with plans to extensively deploy hydrogen stations. Despite Japan maintaining a relatively modest and stable market size of around 100 billion

USD, it leads in terms of the adoption rate of hydrogen infrastructure. By 2050, North America, China and Europe are poised to emerge as the three primary hydrogen markets, collectively commanding approximately 60% of the global market size, followed by Japan and India. Consequently, these nations have implemented specialized policies and strategies to encourage technology revolution and market growth in the hydrogen sector.

## 6. HYDROGEN FOR ENERGY INTEGRATION

One might argue that hydrogen could play a key role in enabling the widespread integration of intermittent renewable electricity into the grid. It provides solutions for addressing both surplus and deficit situations in electricity supply: excess electricity can be transformed into hydrogen through water electrolysis, while hydrogen re-electrification can help compensate for deficits by facilitating electricity time-shifts over longer durations. However, the challenge persists regarding alternative methods for achieving high levels of irregular renewables integration in electricity production, particularly in areas where underground storage of hydrogen is not geologically or economically viable.

An aspect that sets electrolysis apart from other electricity storage technologies, such as batteries, is its versatility. Electrolysis doesn't limit itself to power-to-power applications; instead, it transforms electricity into an energy vector that can be utilized across various applications. However, in the hierarchy of flexibility solutions for integrating renewable energies of fluctuating type, hydrogen is generally not the foremost choice. Nonetheless, it could find its place within the system through alternative means, particularly if recent marketplaces for green hydrogen and electrolysis develop, driven by suitable inducements frameworks. When such electrolysers are not operating at complete capacity, it could serve as a flexibility measure themselves by integrating surplus renewable electricity through additional hydrogen production. In this scenario, the focus shifts from addressing surplus issues to renewable hydrogen production. Ultimately, the competitiveness of hydrogen as a storage option stands to benefit from its deployment as a low-$CO_2$ energy vector in other sectors.

While the remaining storage of electricity alone may not immediately defend the building of hydrogen caverns, such caverns could facilitate various other applications and business opportunities related to hydrogen use in industry and transportation. These may include serving as backup supply and trading hubs, acting as distribution centers, and facilitating hydrogen export/import operations. Additionally, given the fluctuating type of hydrogen generation from excess electricity, integration of hydrogen on a large-scale into the normal gas grid would necessitate buffer capacity to maintain a continuous gas configuration.

## 7. CHALLENGES

Achieving a hydrogen economy poses numerous challenges that must be addressed (Urunkar & Patil, 2023a). These challenges encompass technical issues such as ensuring the constancy and consistency of fuel cells, as well as commercial considerations regarding the costs associated with system integration, infrastructure development, and establishing a supply chain for hydrogen production on a large scale and conveyance. Additionally, commercial support, the establishment of methodological standards, and regulatory frameworks are essential components that require attention. Addressing these challenges demands concerted efforts and endeavors. It's imperative to resolve these issues before widespread adoption of hydrogen technologies can be realized.

The hydrogen economy encompasses a broad array of innovative technologies spanning from major research to practical applications across various stages of production, storage, transportation, and utilization. Both stationary fuel cells and Fuel Cell Vehicles (FCVs) are typical methods for storage and usage hydrogen energy in the realm of fuel cell technology. One avenue of development focuses on polymer electrolyte fuel cell technologies. This entails advancements in noble metal substituted catalyst materials, economic electrolyte materials, as well as additional allied tools to enhance electrode reactions. For industrial applications, there's a growing interest in solid oxide fuel cells, with efforts aimed at increasing power density and operational efficiency. This requires ongoing research into new catalyst materials and enhancing consistency of system for durable use. Moreover, in the realm of generation of hydrogen power at large scale, the megawatt (MW) level, its technical practicality needs further validation. Additionally, specific technologies for hydrogen combustion are anticipated to mitigate the emission of nitrogen oxides (NOx). Continued efforts in research and development are essential to address these technological challenges and drive advancements in the hydrogen economy.

## 7.1 Economic Concerns

The establishment of infrastructure and robust supply chain presents additional critical challenge in realizing the hydrogen based economy. A primary concern regarding the hydrogen economy revolves around its total costs. However, while achieving cost reductions is a key objective, a significant portion of this reduction is expected to stem from economies of scale in the coming times. As volumes of production brought to the market increase, costs of production are anticipated to decrease accordingly. For the widespread adoption of hydrogen technologies on a large scale, it is essential to identify target suppliers and users. This entails introducing a substantial amount of production as well as utilization equipment beforehand. Additionally, infrastructure for storage and transportation of hydrogen, such as pipelines and hydrogen stations is indispensable. The implementation of necessary tools for hydrogen stations or the construction of pipelines, coupled with the popularization of Fuel Cell Vehicles (FCV's) and capacity building, necessitates a substantial initial investment. While this initial investment is unavoidable, it is crucial for laying the foundation for the successful implementation of hydrogen technologies on a broader scale.

Furthermore, achieving a carbon dioxide free supply chain is a critical goal for the hydrogen economy. When utilizing hydrogen as a storage medium to balance renewable power generation, the development of large electrolysis cells with high electrical density becomes paramount. Additionally, in scenarios where hydrogen is produced from natural gas associated with oil or brown coal, integrating Carbon Capture and Storage (CCS) technology is anticipated to mitigate carbon dioxide emissions throughout the life cycle assessment. Moreover, as the hydrogen economy introduces novel aspects to production of energy and usage, like importing from other countries or tapping into unused local resources, there is a pressing need to reassess and update existing standards and regulations. This ensures that the regulatory framework adequately addresses the unique challenges and opportunities presented by hydrogen-based energy systems while promoting sustainability and environmental responsibility.

## 7.2 Current Challenges for Green Hydrogen

Green hydrogen is produced through the process of electrolysis, which involves splitting water using an electric current in an electrolyzer (Cho, 2021). This method employs renewable electricity to generate hydrogen and oxygen, resulting in significantly lower carbon emissions related to fossil fuels as well as grey hydrogen. But, to ensure the production process is genuinely carbon-free, the electricity source must be renewable. Utilizing a conventional power grid for hydrogen production undermines its environmental benefits, as a substantial portion of electricity is often derived from non-renewable sources such as coal, oil, and natural gas (Searle & Zhou, 2021). To establish a completely "green" hydrogen facility, keen renewable energy sources like wind farms or solar plants are required to ensure consistent zero-carbon electricity. However, this approach limits the hydrogen facility capacity factor to that of the primary renewable resource, leading to substantial additional capital expenditures for compressed hydrogen storage and electrolyzers. Moreover, the green hydrogen market is still in its early stages, presenting various challenges. Nevertheless, these challenges also present opportunities for both emerging as well as established companies functioning in the sector of green hydrogen to develop innovative solutions and drive industry growth.

1. The hydrogen economy is anticipated to generate numerous new job openings, yet the slow pace of technical learning and a shortage of requisite skills have resulted in an insufficient pool of specialized labor needed to support this emerging industry. This shortage poses a significant obstacle to the industry's development and maturity. Furthermore, transportation and storage of hydrogen, a extremely explosive and inflammable gas, necessitates significant investments

in specialized carriers as well as pipelines. The considerable uncertainties and expenditures associated with adapting and transforming infrastructure for generation, distribution, and storage systems are among the primary challenges facing the industry. These astronomical costs and uncertainties represent significant barriers to the widespread adoption and advancement of the hydrogen economy (Agnolucci & Mcdowall, 2013).

2. The limited knowledge of optimal design poses challenges to the profitability and stability of organizations aiming to meet market demand for green hydrogen. Enhancing plant designs for green hydrogen generation requires careful optimization of end to end systems, which might be complex and costly due to a lack of market data. Moreover, when plants for generation of green hydrogen are integrated within present industrial groups, the design as well as scaling up process becomes yet more intricate. Special attention must be paid to minimizing any negative commercial impact on existing operations during the transition phase. This complexity adds further challenges to the optimization of plant designs and the successful implementation of green hydrogen systems.
3. Commercializing green hydrogen presents a significant challenge as a result of the distribution and storage stringent requirements. While it is possible to produce green hydrogen economically in regions with abundant sunlight, like Portugal, Spain, Tunisia, and Australia, industrial procurers are often not situated nearby. This necessitates the installation of dedicated transportation infrastructure, leading to increased costs and lead times. Furthermore, the valuation of green hydrogen relies on certification of "Guarantee of Origin" and convertibility for carbon credit. However, both of these approaches are yet in the stage of development and are subject to ongoing arguments. The uncertainty surrounding these frameworks further complicates the monetization of green hydrogen and adds to the challenges faced by the industry.
4. The applications as well as production of green hydrogen incur significant energy losses all over the supply chain. In the process of hydrogen production via electrolysis approximately 30–35% of the energy is lost. Furthermore, converting hydrogen to carriers such as ammonia or liquefying it results in an additional energy loss of 13–25%. Transporting

hydrogen further contributes to a loss of 10–12% (Renewable & Energy Agency, 2020). Finally, the use of hydrogen in fuel cells leads to energy loss of 40 to 50% additionally. These inadequacies highlight the need for improvements throughout the entire green hydrogen supply chain. A considerable amount of renewable energy will be needed to power green hydrogen based electrolyzers, unless such losses are addressed and minimized, in order to compete effectively with end-use electrification. Addressing these inefficiencies is crucial for enhancing the viability and competitiveness of green hydrogen as an energy carrier.

## 8. POLICIES DRIVING THE HYDROGEN REVOLUTION

To realize the green hydrogen revolution, it is crucial to accelerate the deployment of renewable energy sources. Countries endowed with abundant renewable energy resources should prioritize the local production of green hydrogen. This approach not only creates economic opportunities but also enhances energy security by reducing reliance on imported fuels and mitigating exposure to disruptions and price fluctuations in the oil market. For nations having access to inexpensive natural gas, an alternative strategy involves investing in carbon capture technologies to facilitate the blue hydrogen production. This approach allows for the utilization of existing infrastructure and resources while reducing carbon emissions associated with hydrogen production. By leveraging their respective energy assets, countries can play a significant role in advancing the global transition to a hydrogen-based economy.

Governments play a crucial role in supporting the downstream utilization of hydrogen by implementing policies aimed at fostering the adoption of fuel cell electric vehicles (FCEV's) powered by hydrogen and establishing hydrogen refueling stations (HRS's). Moreover, there is a need to enhance the transport and distribution infrastructure to facilitate the widespread use of hydrogen. In cases where domestic demand exceeds internal production capabilities, countries should seek international supply agreements. For instance, a notable agreement was signed between E.ON, largest energy company in Germany and the green power division of Australian miner Fortescue Metals to transport green hydrogen to Europe, thereby reducing reliance on traditional energy sources such as Russia. Furthermore, the developing a hydrogen economy requires the timely training of a sufficient number of skilled workers, which should be supported by national policies. Investing in workforce development is essential for ensuring the successful implementation and expansion of hydrogen technologies across various sectors.

To facilitate the changeover to green hydrogen, especially for the period of the initial stages when costs are high, it is essential to provide financial support to businesses undergoing the changeover. Concessional funds and assistance from development finance institutions, can significantly aid the operation of pioneering green hydrogen projects. This support will accelerate the adoption of green hydrogen technologies, particularly in developing nations where resources may be limited (Sisternes & Jackson, 2020). Figure 3 presents an outline of the essential governing policies necessary to develop the hydrogen economy. These policies are crucial for creating an enabling environment that fosters investment, innovation, and collaboration in the hydrogen sector. By implementing these regulatory measures, governments can promote the growth of the green hydrogen market and drive sustainable development in the energy sector.

*Figure 3. Policy outline for a quicker changeover to low-carbon hydrogen.*

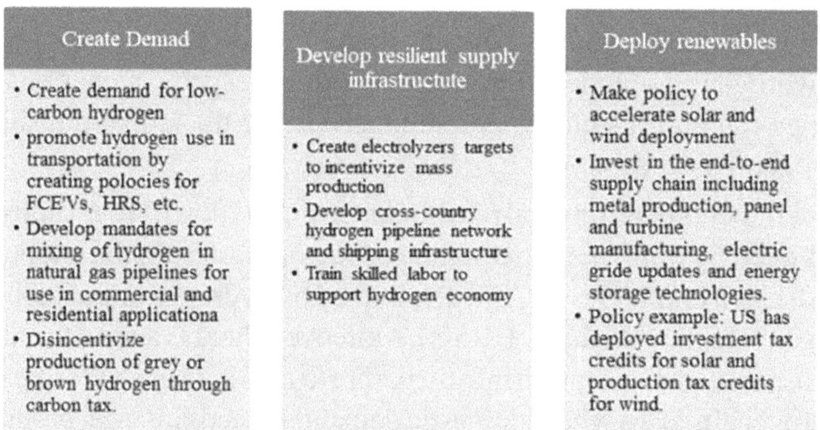

## 9. GLOBAL COLLABORATION FOR HYDROGEN INNOVATION AND SUSTAINABILITY

Recognizing the importance of transitioning to a hydrogen economy, major nations as well as regions have initiated cooperation efforts and negotiations via different platforms and channels to address challenges and overcome barriers. Key platforms for international collaboration comprise the International Partnership for Hydrogen and Fuel Cell in the Economy (IPHE) and the International Energy Agency (IEA). These platforms serve as forums for sharing knowledge, exchanging best practices, and fostering collaboration among nations to advance the development and deployment of hydrogen and fuel cell technologies. Through these cooperative initiatives, countries work together to accelerate the changeover towards a sustainable hydrogen energy economy.

Currently, the International Energy Agency (IEA) serves as a dedicated international organization for cooperation and consultation in the energy sector, concentrating on both policy implementation as well as technology development. The Committee on Energy Research and Technology (CERT) has been formed, among its standing committees, to develop technology roadmaps, investigate technology and market trends and endorse international cooperation on the demonstration, development, and deployment of energy technologies to address sector challenges. International collaboration as well as information exchange is typically facilitated via a series of Implementing Agreements (IAs), also known as "Multilateral Energy Technology Initiatives," succeeded by CERT. Two such agreements specifically focus on hydrogen economy and fuel cell technology:

- Advanced Fuel Cell Implement Agreement (AFCIA)
- Hydrogen Implement Agreement (HIA)

These agreements serve as platforms for collaborative research, development, and deployment efforts aimed at advancing hydrogen and fuel cell technologies on an international scale.The Hydrogen Implement Agreement (HIA) operates under the Renewable Energy Working Party within the International Energy Agency (IEA). Its primary objective is to foster international association for the advancement as well as utilization of hydrogen technology. By starting collective activities within specific sectors, member nations and corporations participate in information sharing and collaborate on the development of implementation plans. Additionally, the Advanced Fuel Cell Implement Agreement (AFCIA) operates under the End Use Working Party, focusing on sharing knowledge and exchanging information specifically related to advanced fuel cell technologies. Through these agreements, stakeholders collaborate to drive innovation and progress in hydrogen and fuel cell technologies on an international level.

On the contrary, the International Partnership for Hydrogen and Fuel Cells in the Economy (IPHE) was founded in 2003 with the objective of advancing the adoption of hydrogen fuel cells via intergovernmental collaboration. IPHE's primary focus is on advocating for the research and development as well as utilization of hydrogen fuel cells. Member governments and institutions participate by sharing updates on policy and development progressions through Country Updates and seek potential market opportunities and partnerships during steering committee meetings. Additionally, IPHE's working groups engage in activities related to regulations, standards, and references, as well as provide support for educational initiatives. Through these efforts, IPHE aims to accelerate the global transition to hydrogen fuel cells.

In addition to these institutions and platforms, many significant international conferences have arisen in the field of hydrogen economy. These include events such as the World Hydrogen Technologies Conventions and the World Hydrogen Energy Conference organized by the International Association for Hydrogen Energy, the International Conference on Hydrogen Safety (ICHS) organized by the non-profit organization International Association HySafe and the Fuel Cell Seminar & Energy Exposition held in America. These conferences serve as important forums for experts, researchers, industry professionals, and policymakers to exchange knowledge, discuss recent advancements, and collaborate on key issues associated to hydrogen energy and fuel cell technologies.

## 10. CONCLUSION

In conclusion, the journey towards a hydrogen-based economy presents both remarkable opportunities and formidable hurdles. The potential of hydrogen as a clean and versatile energy carrier is undeniable, offering solutions to decarbonize various sectors and mitigate the impacts of climate change. However, transitioning to a hydrogen economy requires concerted efforts from governments, industries, and stakeholders worldwide. Opportunities abound in the form of technological advancements, policy frameworks, and international collaborations. Developments in electrolysis, fuel cell technology, and hydrogen infrastructure pave the way for scalable and sustainable hydrogen production, distribution, and utilization. Moreover, initiatives such as the International Energy Agency (IEA) and the International Partnership for Hydrogen and Fuel Cell in the Economy (IPHE) facilitate global cooperation and knowledge sharing, fostering innovation and progress in the field. However, substantial challenges must be overcome to realize the full potential of a hydrogen-based economy. Technical hurdles, including the efficiency and reliability of fuel cells, as well as the cost-effectiveness of hydrogen production and storage, remain key areas of focus. Additionally, the establishment of robust supply chains, supportive regulatory frameworks, and skilled workforce are essential for the widespread adoption of hydrogen technologies. Despite these challenges, the momentum towards a hydrogen economy continues to grow. With continued research, investment, and collaboration, the vision of a sustainable future powered by hydrogen can become a reality. Navigating the path to a hydrogen-based economy requires diligence, innovation, and collective action, but the rewards in terms of cleaner air, energy security, and economic prosperity are well worth the effort.

## REFERENCES


Agnolucci, P., & Mcdowall, W. (2013). Designing future hydrogen infrastructure : Insights from analysis at different spatial scales. *International Journal of Hydrogen Energy*, *38*(13), 5181–5191. DOI: 10.1016/j.ijhydene.2013.02.042

Ball, M., Weeda, M., Ball, M., & Weeda, M. (2015). The hydrogen economy — Vision or reality? *International Journal of Hydrogen Energy*, *40*(25), 7903–7919. DOI: 10.1016/j.ijhydene.2015.04.032

Bockris, J. O. M. (2013). The hydrogen economy : Its history. *International Journal of Hydrogen Energy*, *38*(6), 2579–2588. DOI: 10.1016/j.ijhydene.2012.12.026

Brandon, N. P., & Kurban, Z. (2017). Clean energy and the hydrogen economy. *Philosophical Transactions. Series A, Mathematical, Physical, and Engineering Sciences*, *375*(2098), 20160400. DOI: 10.1098/rsta.2016.0400 PMID: 28607181

Bridgeland, R., Chapman, A., Mclellan, B., Sofronis, P., & Fujii, Y. (2022). Challenges toward achieving a successful hydrogen economy in the US : Potential end-use and infrastructure analysis to the year 2100. *Cleaner Production Letters*, *3*, 100012. DOI: 10.1016/j.clpl.2022.100012

Cho, R. (2021). *Why We Need Green Hydrogen?*https://news.climate.columbia.edu/2021/01/07/need-green-hydrogen/

Environmental Protection Agency. U. S. (2023). *Greenhouse Gases What are the trends in greenhouse gas emissions and concentrations and health and the environment?*https://www.epa.gov/report-environment/greenhouse-gases

Gregory, O., Ng, D., & Long, G. (1972). The Electrochemistry of Cleaner Environments. In *Surface Electrochemistry* (pp. 927–976).

Hart, D., Howes, J., Madden, B., & Boyd, E. (2016). *Hydrogen and Fuel Cells: Opportunities for Growth.*

International Energy Agency. (2023). *World Energy Outlook 2023 | Enhanced Reader.* https://www.iea.org/news/the-energy-world-is-set-to-change-significantly-by-2030-based-on-today-s-policy-settings-alone

Jawerth, N. (2022). *What is the Clean Energy Transition and How Does Nuclear Power Fit In?*https://www.iaea.org/bulletin/what-is-the-clean-energy-transition-and-how-does-nuclear-power-fit-in

Kochanek, E. (2022). The Role of Hydrogen in the Visegrad Group Approach to Energy Transition. *Energies*, *15*(19), 7235. https://doi.org/. DOI: 10.3390/en15197235

Lamb, W. F., Wiedmann, T., Pongratz, J., Andrew, R., Crippa, M., Olivier, J. G. J., Wiedenhofer, D., Mattioli, G., Al Khourdajie, A., & House, J. (2021). A review of trends and drivers of greenhouse gas emissions by sector from 1990 to 2018. *Environmental Research Letters*, *16*(7), 073005. https://doi.org/https://doi.org/10.1088/1748-9326/abee4e. DOI: 10.1088/1748-9326/abee4e

Lu, W. (2017). Greenhouse Gas Emissions, Energy Consumption and Economic Growth : A Panel Cointegration Analysis for 16 Asian Countries. *International Journal of Environmental Research and Public Health*, *14*(11), 1436. DOI: 10.3390/ijerph14111436 PMID: 29165399

Mcdowall, W. (2012). Technology roadmaps for transition management : The case of hydrogen energy. *Technological Forecasting and Social Change*, *79*(3), 530–542. DOI: 10.1016/j.techfore.2011.10.002

Mcdowall, W. (2014). Exploring possible transition pathways for hydrogen energy : A hybrid approach using socio-technical scenarios and energy system modelling. *Futures*, *63*, 1–14. DOI: 10.1016/j.futures.2014.07.004

Nikkei. (2013). *Overview of World Hydrogen Infrastructure Projects.*

Noussan, M., Raimondi, P. P., Scita, R., & Hafner, M. (2020). The Role of Green and Blue Hydrogen in the Energy Transition - A Technological and Geopolitical Perspective. *Sustainability (Basel)*, *13*(1), 298. DOI: 10.3390/su13010298

Ren, X., Dong, L., Xu, D., & Hu, B. (2020). Challenges towards hydrogen economy in China. *International Journal of Hydrogen Energy*, *45*(59), 34326–34345. DOI: 10.1016/j.ijhydene.2020.01.163

Renewable, I., & Energy Agency. (2020). *Green Hydrogen: A guide to policy making*.

Rosen, M. A., & Koohi-fayegh, S. (2016). The prospects for hydrogen as an energy carrier : An overview of hydrogen energy and hydrogen energy systems. *Energy, Ecology & Environment*, *1*(1), 10–29. DOI: 10.1007/s40974-016-0005-z

Satyapal, S. (2017). *Hydrogen: A Clean, Flexible Energy Carrier*. https://www.energy.gov/eere/articles/hydrogenclean-flexible-energy-carrier

Searle, S., & Zhou, Y. (2021). *Don't let the industry greenwash green hydrogen*.

Sisternes, F. de, & Jackson, C. (2020). *Green Hydrogen in develoing contries*.

Urunkar, R. U., & Patil, S. D. (2021a). Enhancement of heat and mass transfer characteristics of metal hydride reactor for hydrogen storage using various nanofluids. *International Journal of Hydrogen Energy*, *46*(37), 19486–19497. DOI: 10.1016/j.ijhydene.2021.03.090

Urunkar, R. U., & Patil, S. D. (2021b). Numerical study on performance evaluation of multi-tubular sodium alanate hydride reactor by enhancing heat and mass transfer characteristics using nanofluids. *Proceedings of the 26th National and 4th International ISHMT-ASTFE Heat and Mass Transfer ConferenceDecember 17-20, 2021, IIT Madras, Chennai-600036, Tamil Nadu, India*, 2613–2618. DOI: 10.1615/IHMTC-2021.3950

Urunkar, R. U., & Patil, S. D. (2023a). Hydrogen Storage Technologies and Related Heat and Mass Transfer Studies. In *Hydrogen Fuel Cell Technology for Mobile Applications* (pp. 182–206). IGI Global., DOI: 10.4018/978-1-6684-6721-3.ch008

Urunkar, R. U., & Patil, S. D. (2023b). Performance analysis of sodium alanate hydride reactor with different nanofluids. *International Journal of Hydrogen Energy, 48*(96), 37944–37953. DOI: 10.1016/j.ijhydene.2023.02.105

Yue, M., Lambert, H., Pahon, E., Roche, R., Jemei, S., & Hissel, D. (2021). Hydrogen energy systems: A critical review of technologies, applications, trends and challenges. *Renewable & Sustainable Energy Reviews, 146*, 111180. DOI: 10.1016/j.rser.2021.111180

# Chapter 3
# Progress and Obstacles in Green Hydrogen Production

**Sankalp Arpit**
https://orcid.org/0000-0001-9669-6978
*Swaniti Initiative, India*

**Tolga Taner**
https://orcid.org/0000-0002-3065-1942
*Aksaray University, Turkey*

## ABSTRACT

*Green hydrogen, which is produced from renewable energy sources has emerged as a significant alternative for the purpose of decarbonizing problematic industries and reducing emissions of greenhouse gases. The purpose of this chapter is to investigate recent developments in the generation of environmentally friendly hydrogen through the utilization of a wide range of technologies, such as solar, wind, ocean thermal energy conversion systems and other energy sources. This chapter provides an in-depth analysis of the primary policies that have been implemented by various countries. In this study, we investigate the current trends in research, the important obstacles that have been encountered, and the future prospects. We emphasize the importance of continuous innovation and collaboration in order to fully realize the potential of green hydrogen in the context of*

DOI: 10.4018/979-8-3693-0827-1.ch003

*the global energy transition. A shift to hydrogen is a significant energy problem that should be adopted across a variety of industries in the future, according to this study, which highlights the importance of this transition.*

## 1 INTRODUCTION

While steam generated from the direct and indirect combustion of fossil fuels has significantly influenced humanity, its improper utilization has also led to environmental deterioration. Direct combustion encompasses coal-fired steam power plants (Arpit et al., 2023), whereas indirect combustion involves utilizing flue gas for steam generation, such as through heat recovery steam generators (HRSG) (Arpit & Das, 2023). The predicament is exacerbated by the exponential population expansion and the poor utilization of fossil fuels. Diverse writers have proposed numerous strategies to alleviate the impacts of global warming, including renewable energy sources, carbon capture, utilization and storage (CCUS), and bioenergy. Nonetheless, each energy source possesses advantages and disadvantages; for instance, renewable energy encounters intermittency issues, carbon capture, utilization, and storage (CCUS) reduce power plant efficiency when implemented, and bioenergy is significantly water intensive. Transitioning from fossil fuel-based power plants is an urgent necessity. To expedite this pace in the endeavor to save our planet, hydrogen has been the focus of investigation.

Hydrogen is the lightest element, and its utilization generates no pollution, producing solely pure water. A further benefit of utilizing hydrogen is its elevated specific heat capacity. Additionally, the methods of hydrogen production differ. It may be derived from petroleum, natural gas, biofuels, among other sources. Recently, hydrogen created from the electrolysis of water has garnered significant attention. The impetus for this movement is twofold: the minimum or negligible emission of carbon dioxide and the ample availability of the raw material, namely water. Table 1 (Soni & Reddy, 2024) presents a comparison of carbon dioxide emissions from various ways.

*Table 1. A comparison of carbon dioxide production in different hydrogen production methods.*

| **Hydrogen Production Methods** | **$CO_2$ Production** |
|---|---|
| **Water electrolysis with renewable energy** | **0 kg/kg. H2** |
| **Steam reforming with carbon capture** | **9-12 kg/kg. H2** |
| **Steam reforming without carbon capture** | **16-18 kg/kg. H2** |
| **Gasification of coal & HC-rich feedstock** | **19-25 kg/kg. H2** |

Table 2 indicates that hydrogen generated via water electrolysis results in the lowest CO2 emissions, and nowadays, this hydrogen generation technology has acquired considerable traction. Recognizing the significance of green hydrogen, numerous governments have developed their individual strategies. A comparison in this context is illustrated in Table 2 (Jindal et al., 2024).

*Table 2. Different policies around the world for developing a green hydrogen hub.*

| **Country/Region** | **Policy/Strategy** |
|---|---|
| India | National Hydrogen Mission |
| European Union | EU Hydrogen Strategy |
| United States | Hydrogen Energy Earthshot |
| Japan | Hydrogen Roadmap |
| South Korea | Hydrogen Economy Roadmap |

It is undeniable that green hydrogen has gained significant momentum, supported by numerous governments. Additionally, it possesses notable attributes such as the capacity to address climate change, decarbonize challenging sectors, and foster economic job creation and growth. Utilizing this clean and adaptable energy source can expedite the worldwide transition to a low-carbon future while promoting a more resilient, sustainable, and economically robust world.

This study article seeks to provide a comprehensive examination of green hydrogen and its significance in promoting a sustainable energy future. Additionally, it explores several aspects of green hydrogen, encompassing production technologies, sectoral applications, environmental and socio-economic benefits, along with the principal hurdles and barriers impeding its widespread acceptance. This research paper provides an in-depth analysis of green hydrogen's significance in the global energy transition, contributing to the growing knowledge of sustainable energy solutions and encouraging ongoing research, development, and innovation in green hydrogen technologies.

## 2 HYDROGEN PRODUCTION TECHNOLOGY

Hydrogen production technologies are crucial for enabling a hydrogen-centric energy economy and advancing the global shift towards greener energy, attracting considerable international focus. The principal techniques of hydrogen production are fossil fuel reformation, water electrolysis, and biological processes (Tuluhong et al., 2024). Among these approaches, steam reforming natural gas is the primary technique due to its economic efficiency. Nonetheless, it produces significant carbon dioxide emissions, which compromises the objective of reducing greenhouse gases. Coal gasification is a pivotal technology to produce green hydrogen (Burchart et al., 2023). Biological hydrogen production utilizes microorganisms, including algae and bacteria, to degrade organic matter and generate hydrogen. It is ecologically sustainable, although limited by insufficient efficiency. In contrast, water electrolysis offers the potential for generating zero-emission hydrogen when powered by renewable energy sources. The advancement and application of hydrogen technologies are widely acknowledged as essential for achieving sustainable development.

Various researchers have examined this field, producing green hydrogen from renewable energy sources including sun and wind, utilizing different configurations and technologies (Ishaq & Dincer, 2021). Several of them have also diversified into the biomass business. A collection of papers has been compiled and is presented in Table 3. Despite the list's comprehensiveness, an effort may be undertaken to assemble certain pieces from it.

Table 3 specifies technology, location, and the concluding observation. The technologies encompass sun, wind, and hydropower. The locations include Türkiye, India, Australia, Nepal, Algeria, the Sahara, and Taiwan.

The final observation includes the responsiveness of financial metrics to the cost of green hydrogen, various policies for the extensive adoption of green hydrogen in India, the capacity of hydropower to produce green hydrogen through the hydropower method, a techno-economic assessment of green hydrogen production using concentrating solar thermal technology (CST), a comparative techno-economic analysis of two technologies (parabolic trough collector and parabolic dish collector), a hydrogen roadmap for Türkiye utilizing solar energy, a techno-economic evaluation of green hydrogen production via a PV/PEM electrolyser system, the development of a hydrogen roadmap in the Sahara region, and the creation of a green hydrogen roadmap for Taiwan employing wind energy.

*Table 3. A collection of reviews dealing with renewable sources of energy for green hydrogen production*

| Technology | Place | Conclusive Remark | Reference |
|---|---|---|---|
| Hydropower | Nepal | Potential of hydropower to produce green hydrogen through the hydropower route. | (Thapa et al., 2021) |
| Solar | Australia | Sensitivity of financial parameters to the cost of green hydrogen. | (Rezaei et al., 2024) |
| Solar | India | Various policies for the mass-scale adoption of green hydrogen in India | (Jindal et al., 2024) |
| Solar | India | Techno-economic analysis of green hydrogen production through concentrating solar thermal technology (CST). | (Kumar & Kumar, 2022) |
| Solar | India | Technoeconomic analysis comparison of two technologies, i.e. Parabolic Trough Collector and Parabolic Dish Collector. | (Soni & Reddy, 2024) |
| Solar | Türkiye | Hydrogen roadmap for Turkey through solar energy. | (Karayel et al., 2022) |
| Solar | Algeria | Techno-economic assessment of green hydrogen production through PV/PEM electrolyser system. | (Ghribi et al., 2013) |
| Solar | Sahara | Development of a hydrogen road map in the Sahara region. | (Mohamed et al., 2016) |
| Wind | Taiwan | Development of a green hydrogen roadmap for Taiwan through wind energy | (Ding et al., 2024) |

There is a wide variety of green hydrogen synthesis technologies available, in addition to solar and wind power. The generation of environmentally friendly hydrogen through the utilization of an ocean-thermal energy conversion system was proposed by one of the authors (Ahmadi et al., 2013). Several authors have proposed the utilization of a hybrid energy system to generate environmentally friendly hydrogen (Ding et al., 2024). In this work, the author has optimized the arrangement of pumped hydro with floating solar using optimization.

Several research evaluations have demonstrated that the production of green hydrogen through the utilization of renewable energy sources is not only feasible but also feasible. However, the general acceptance of this product has been hampered by several factors, including the increased costs involved. A decrease in the Levelized Cost of Hydrogen (LCOH) is the outcome of growing demand, which is in accordance with the principle of economies of scale (Arpit et al., 2024). Before deciding about this technology, it is necessary to first determine whether there is a need within the global supply chain.

## 3 CHALLENGES

To realize a hydrogen economy, it is necessary to successfully overcome a number of significant challenges. This encompasses both technical challenges, such as ensuring the dependability and stability of renewable energy sources, and financial concerns, which are related to the substantial expenses associated with system integration, infrastructure development, and the establishment of a comprehensive hydrogen production and distribution network. Technical challenges include this. In addition to this, it is vital to have strong economic support, the establishment of established methodologies, and the enforcement of explicit regulatory frameworks. In order to address these challenges, it would be necessary to make efforts that are cooperative and continuing. Prior to the widespread use of hydrogen technology, it is imperative that these concerns be addressed initially.

The hydrogen economy involves a wide variety of cutting-edge technologies, ranging from fundamental research to practical applications along the entire value chain, which includes production, storage, transportation, and usage of hydrogen. Typically, hydrogen is stored and consumed in fuel cells that are stationary, as well as in fuel cell vehicles (FCVs), which are also known as fuel cell applications. The development of polymer electrolyte fuel cells is an important area of research and development. These fuel cells have an emphasis on the improvement of catalyst materials that can replace noble metals, the production of electrolytes that are inexpensive, and the optimization of related technologies to boost the efficiency of electrode reactions. Due to the fact that solid oxide fuel cells can increase power output and operating efficiency, they are becoming increasingly popular in industrial settings. For these systems to continue to advance, there must be constant research into innovative catalysts and advances in system stability to ensure long-term reliability. Furthermore, additional validation of the technological feasibility is necessary to generate a significant amount of hydrogen electricity at the megawatt scale. To overcome these technological challenges and encouraging the development of the hydrogen economy, ongoing research and development are very necessary.

## 4 CONCLUSION

The shift to an economy based on hydrogen brings with it a tremendous deal of potential but also presents several significant challenges. As a clean and versatile energy carrier, hydrogen has a huge potential to decarbonize multiple industries and contribute to the reduction of climate change. Hydrogen possesses significant potential for these purposes. Recognizing this potential, on the other hand, calls for coordinated efforts from governments, companies, and stakeholders all around the world. Technological advancements, rules that are supportive of them, and international cooperation have all contributed to the existence of numerous opportunities. Methods of production, delivery, and usage are becoming more scalable and environmentally friendly as a result of developments in electrolysis, fuel cell systems, and hydrogen infrastructure.

For the purpose of fostering collaboration and accelerating innovation, it is vital to have global efforts, such as those that are being led by the International Energy Agency (IEA) and the International Partnership for Hydrogen and Fuel Cells in the Economy (IPHE). However, in order to fully exploit the promise of hydrogen, there are other important obstacles that need to be overcome. Enhancing the effectiveness and dependability of fuel cells, lowering the costs associated with hydrogen production and storage, and developing reliable supply systems are the primary challenges that need to be addressed. The establishment of comprehensive legal frameworks and the development of a skilled workforce are both necessary steps toward achieving widespread acceptance.

In spite of these obstacles, continued progress is being made toward the establishment of a hydrogen-based economy. The goal of a future powered by hydrogen that is both clean and environmentally friendly is possible if diligent study, strategic investment, and international cooperation are all put into practice. In spite of the fact that the realization of this vision requires endurance, ingenuity, and community effort, the project is one that merits dedication because of the long-term benefits it offers, which include enhanced air quality, increased energy security, and economic growth.

Hydrogen technologies, which are considered to be the energy of the future, are getting more and more applications and are becoming more appropriate. After the problem with supply security is handled, it is projected that it will emerge as the most important and advantageous alternative energy source in the future, emerging in a manner that is more effective and efficient. It is projected that once it has been fully implemented, it will be the component that is the most efficient in terms of delivering renewable energy within the ecosystem of the entire world.

## ACKNOWLEDGEMENT

The authors would like to thank the scientists and authors of the previous works cited in this book chapter. This book chapter will be a source of inspiration for academia, students and industry. Authors also acknowledge support from IGI Global Scientific Publishing.

## REFERENCES

Ahmadi, P., Dincer, I., & Rosen, M. A. (2013). Energy and exergy analyses of hydrogen production via solar-boosted ocean thermal energy conversion and PEM electrolysis. *International Journal of Hydrogen Energy*, *38*(4), 1795–1805. DOI: 10.1016/j.ijhydene.2012.11.025

Arpit, S., & Das, P. K. (2023). A state-of-the-art review of heat recovery steam generators and waste heat boilers. *Energy Efficiency*, *16*(8), 99. Advance online publication. DOI: 10.1007/s12053-023-10179-5

Arpit, S., Kumar, P., Das, P. K., & Dash, S. K. (2023). Application of exergy analysis in understanding the performance of a coal-fired steam power plant (120 MW) with single reheat and regenerative configuration. *Journal of Thermal Engineering*, *9*(2), 497–509. DOI: 10.18186/thermal.1285229

Arpit, S., Pai, S., & Choudhary, S. M. (2024). Should state-owned coal companies harness green hydrogen? https://www.swaniti.com/wp-content/uploads/2024/10/Green-hydrogen-final.pdf

Burchart, D., Gazda-Grzywacz, M., Grzywacz, P., Burmistrz, P., & Zarębska, K. (2023). Life Cycle Assessment of Hydrogen Production from Coal Gasification as an Alternative Transport Fuel. *Energies*, *16*(1), 383. Advance online publication. DOI: 10.3390/en16010383

Ding, J. W., Fu, Y. S., & Lisa Hsieh, I. Y. (2024). The cost of green: Analyzing the economic feasibility of hydrogen production from offshore wind power. Energy Conversion and Management: X, 24, 100770. DOI: 10.1016/j.ecmx.2024.100770

Ghribi, D., Khelifa, A., Diaf, S., & Belhamel, M. (2013). Study of hydrogen production system by using PV solar energy and PEM electrolyser in Algeria. *International Journal of Hydrogen Energy*, *38*(20), 8480–8490. DOI: 10.1016/j.ijhydene.2012.09.175

Ishaq, H., & Dincer, I. (2021). Comparative assessment of renewable energy-based hydrogen production methods. *Renewable & Sustainable Energy Reviews*, *135*, 110192. DOI: 10.1016/j.rser.2020.110192

Jindal, A., Shrimali, G., & Tiwary, N. (2024). At scale adoption of Green Hydrogen in Indian Industry: Costs, subsidies and policies. *Energy for Sustainable Development*, *83*, 101549. DOI: 10.1016/j.esd.2024.101549

Karayel, G. K., Javani, N., & Dincer, I. (2022). Green hydrogen production potential for Turkey with solar energy. *International Journal of Hydrogen Energy*, *47*(45), 19354–19364. DOI: 10.1016/j.ijhydene.2021.10.240

Kumar, S., & Kumar, K. R. (2022). Techno economic feasibility study on hydrogen production using concentrating solar thermal technology in India. *International Journal of Hydrogen Energy*, *47*(89), 37708–37723. DOI: 10.1016/j.ijhydene.2022.08.285

Mohamed, B., Ali, B., Ahmed, B., Ahmed, B., Salah, L., & Rachid, D. (2016). Study of hydrogen production by solar energy as tool of storing and utilization renewable energy for the desert areas. *International Journal of Hydrogen Energy*, *41*(45), 20788–20806. DOI: 10.1016/j.ijhydene.2016.07.034

Rezaei, M., Akimov, A., & Gray, E. M. A. (2024). Cost-competitiveness of green hydrogen and its sensitivity to major financial and technical variables. *International Journal of Hydrogen Energy*, *90*, 1365–1377. DOI: 10.1016/j.ijhydene.2024.10.068

Soni, N., & Reddy, K. S. (2024). Techno-economic assessment of concentrated solar power technologies integrated with thermal energy storage system for green hydrogen production. *International Journal of Hydrogen Energy*, *72*, 1184–1203. DOI: 10.1016/j.ijhydene.2024.05.409

Thapa, B. S., Neupane, B., Yang, H., & Lee, Y.-H. (2021). Green hydrogen potentials from surplus hydro energy in Nepal. *International Journal of Hydrogen Energy*, *46*(43), 22256–22267. DOI: 10.1016/j.ijhydene.2021.04.096

Tuluhong, A., Chang, Q., Xie, L., Xu, Z., & Song, T. (2024). Current Status of Green Hydrogen Production Technology: A Review. In Sustainability (Switzerland) 16(20). DOI: 10.3390/su16209070

Section 2

# Application of Hydrogen Technology in Automotive and Industrial Sectors

Chapter 4

# Green Hydrogen for Cement Production:

## A Decarbonization Pathway With Hybrid Renewable Energy Sources

**Parijat Rai**

*Department of Electrical and Electronics Engineering, SRM Institute of Science and Technology, Modinagar, India*

**Dharmbir Prasad**

https://orcid.org/0000-0002-9010-9717

*Department of Electrical and Electronics Engineering, SRM Institute of Science and Technology, Modinagar, India*

**Rudra Pratap Singh**

https://orcid.org/0000-0001-7352-855X

*Department of Electrical Engineering, Asansol Engineering College, Asansol, India*

**Ranjeet Singh**

*Department of Electrical Engineering, Faculty of Technology, Delhi University, India*

**Tanuja**

*Department of Electrical Engineering, Asansol Engineering College, Asansol, India*

**Ranadip Roy**

https://orcid.org/0000-0003-2111-2581

*Department of Electrical Engineering, Sanaka Educational Trust's Group of Institutions, Durgapur, India*

DOI: 10.4018/979-8-3693-0827-1.ch004

**Md. Irfan Khan**

*Marketing Department, Sawaria Pipes Pvt. Ltd., Hyderabad, India*

## ABSTRACT

*The cement industry is essential to the growth and development of the Indian economy and infrastructure due to the enormous rise in the housing sector. Any rapidly developing economy faces a lot of energy demand and environmental issues. Green hydrogen is generated through water electrolysis using renewable sources solar, wind, etc playing an important role in resisting climate change. This work describes the implementation of solar energy for the calcination process in cement production. The proposed model have PV power along with battery storage as a key source while diesel generator as back up for emergency requirements. The proposed Chhattisgarh site is located at coordinates of latitude: 21o16.7 N and longitude: 81o52.0 E. This study suggests PV system capacity factor is 18.4% with 4,361 hours/year of operation. The total production of solar energy is 24,234,917 kWh/year. This solar power is meeting the green hydrogen needs of proposed cement plant with 325 kg/year.*

## INTRODUCTION

The advancement of technology has expedited the industrial sector's innovation and productivity since the industrial revolution. The need for resources is also expanding quickly in the meanwhile. Fossil fuels provide for a sizable portion of the thermal energy needed for cement manufacture. The most widely used fossil fuels are biomass, natural gas, coal and petro coke. Additionally, the burning of fossil fuels emits a number of harmful substances into the atmosphere. As a result, it's crucial to cut back on fossil fuel use while still meeting the thermal energy needs of the cement industry. Numerous dangerous gases are released into the atmosphere by the cement industry. According to (Sahoo et. al., 2023), this industry contributes around 13% and 8% of the world's total greenhouse gas emissions and anthropogenic carbon dioxide to the environment, respectively.

According to estimates, the manufacture of one ton of clinker contributes 0.9 to one ton of $CO_2$ to the atmosphere (Sahoo et. al., 2023). CaCO3 breakdown is responsible for around half of these emissions, with the remaining emissions coming from fuel combustion, chemical processes involving salts and oxides, etc. The presence of carbonates in the raw material causes process emissions. The majority of cement CO2 emissions (almost 70% in the most efficient cement plants) are produced during the calcination of carbonates, particularly calcium carbonate, which makes up around 60% of the clinker raw material (Sousa et. al., 2023).

One of the industries that consume bulk electricity is the cement industry, which relies heavily on fossil fuels for its thermal energy needs. The methods (dry/wet) used to create cement have a significant impact on the quantity of thermal energy needed. For instance, 3200–5500 MJ of thermal energy are required to produce one ton of clinker formed cement (Sahoo et. al., 2023). In comparison to pre-industrial levels, anthropogenic emissions have already resulted in an increase in global temperature of 1.1°C. Most country agree that achieving net zero by 2050 is necessary to boost the likelihood of keeping this temperature increase to 1.5°C or less. Due to this renewed attention, emissions from all energy end users must be reduced. While energy efficiency, electrification, and renewable energy sources can provide 70% of the necessary mitigation, hydrogen will be required to decarbonize end uses such as heavy industry and long-distance transportation (IREA, 2022). Due public awareness to combat climate change and as a result of ongoing price reductions for RE sources, improvements in energy storage technology and various decarbonization strategies used in the cement industry, the energy shift is increasing (d'Hubert et. al., 2021). A sustainable framework must be built to guarantee an energy-efficient production process (Pei et. al., 2021). Dihydrogen (H2), sometimes known as hydrogen, is becoming more widely accepted by different industries as a safe and dependable energy source for decarbonization and defossilization. By 2024, the expected growth in the world's hydrogen demand from 70 million tonnes in 2019 to 120 million tonnes. The seventh UN goal of 'affordable and clean energy' may also be achieved through hydrogen development (Osman et. al., 2021).

The use of hydrogen in various processes within the cement industry can vary widely depending on the specific technology, equipment and goals of emissions reduction. Following are the key applications of hydrogen in cement plant (d'Hubert et. al., 2021):

*(i). Hydrogen as a reducing agent in kilns:* Substituting hydrogen for a portion of the fossil fuels used in the kiln process might typically involve hydrogen in the range of 10% to 30%. However, this can vary significantly based on the specific kiln design and technology.
*(ii). Hydrogen for power generation:* The amount of hydrogen in the electricity generation mix would depend on the source and availability of hydrogen. In some cases, it might be a small amount of the overall energy mix, around 5% to 20%.
*(iii). Hydrogen in clinker production:* The use of hydrogen as a reducing agent in clinker production is still in the experimental stage. Depending on the technology used, hydrogen requirement might range from 10% to 50% or more.
*(iv). Carbon capture with hydrogen:* In CCU processes, hydrogen can be used to capture $CO_2$ emissions. The amount of hydrogen used can vary but is typically around 10% to 20% in the gas stream used for capture.

India a developing country needs a shift from fossil fuel to other environmental clean option such as green hydrogen produced by RE sources through the electrolysis of water' is a positive to approach the challenges posed by climate change and need to shift our energy system. Additionally, the government has mandated that for every kilogram of hydrogen produced, carbon emissions must not exceed two kilograms of carbon dioxide equivalent. The government has also developed incentive programs to promote the creation of electrolyzers and green hydrogen. According to a notification, from June, out of the $17,490 allotted for the Strategic Interventions for 'Green Hydrogen Transition Mission', $13,050 would be used to encourage the generation of green hydrogen and the remaining $4,440 would be used to produce electrolyzers (Live Mint, 2023). This coordination also enables electrolysers to add flexibility to the grid, completing other smart electrification options like batteries and demand response. The present study focused opportunities and difficulties of green hydrogen in cement industry. By replacing fossil fuels with green hydrogen, the pollutant emission can be reduced to a large extent.

The remainder of the work is organized as follows: Problem Statement, Review of Research Background, Mathematical Modeling, Energy Potential Assessment, System Modeling with Solution Approach, Results and Discussion. Finally, the research has come to an end with conclusive remarks.

## PROBLEM STATEMENT

The cement industry is India's second most important primary and basic sector, after only the iron and steel industry in importance. Cement is required for the construction of various forms of infrastructural facilities such as roads, bridges, dams, buildings, canals and so on. Cement production was the basis for the design of the hybrid RE powered cement plant will curtail massive pollutant emission. In this study following are the major objective have been covered towards green manufacturing:

- Analyze the sensitivity of a design variable by performing parametric simulations
- Minimize expected energy costs by optimizing the model parameters
- Determine the potential to reduce $CO_2$ emissions
- Assess the solar model's economic viability

## RESEARCH BACKGROUND REVIEW

The usage of hydrogen in cement production is an evolving field and research and development efforts are ongoing to optimize the processes and achieve greater emissions reductions. Following are the recent researches suggesting various solutions for green industrial production:

*(i). Green hydrogen application in cement plant:* Recently, a study highlighted the possibility for energy collaboration between Europe and North Africa that would be advantageous to both parties, with

solar power and renewable hydrogen generation in cost-effective way (Van Der Zwaan et al., 2021). In addition, the reliance on variables like the season, time of day and network location in different regions, provides an integrated model that combines green hydrogen injection into electricity and gas networks (Saedi et al., 2021). Further study, discussed the economic worth of PtG installations based on RE sources harvesting (Li &Mulder, 2021). It recognizes that these installations now have a negative economic value but makes the case that future cost reductions and efficiency gains could increase their viability (Dey et al., 2022). Next, a research suggested retrofitting method for decarbonizing coal-fired power plants that uses only RE sources and flexible hydrogen generation, highlighting the need for pilot projects in wind-rich locations (Hassan et al., 2023). Furthermore, the Clean-India objective is emphasized need for institutional and regulatory changes for commercially accessible technology (Abhyankar et. al., 2023). Then, a multigenerational energy system based on solar and biomass energy is examined in another research, which reveals important advantages in terms of increased power output and decreased CO2 emissions (Assareh et. al., 2023). However, a combined strategy for ZEVs energy demand have requirements of green energy sources that are renewable for sustainable transportation and biofuels can achieve UNFCCC and SDG 13 goals in the long run (Sandaka et. al., 2023).

*(ii). Advancement in water electrolysis for hydrogen:* The study highlighted the growing relevance of water electrolysis technologies for producing green hydrogen within the framework of overall decarbonization efforts (Kumar& Lim, 2022). Then, another study emphasized the difficulties and shortcomings of various RE sources, including wind, solar and hydroelectric power (Maciel et al., 2023). Further, a study presented GH2 mobile net, a potential deep learning method for estimating the generation of green hydrogen from mixes of organic waste using picture data (Torky et. al., 2023). Though, further developments in this field, including models for various waste materials. Thus, biomass gasification is considered a viable substitute for fossil fuels since it can produce low-carbon hydrogen from biomass as suggested in (Pocha et. al., 2023).

*(iii). Renewable energy integration:* In the context of the global energy transition and sustainability objectives, (Ruhnau, 2022) highlights the rising significance of RE sources and hydrogen generation. Then, a study emphasizes the synergistic benefits of flexible hydrogen production and variable RE sources to encourage investments in green way of electrolysis (Ji, 2022). Further, Ji et al., (2022) have suggested exclusion of electrolyzer from electricity price enhancements in order to encourage investment, policy actions are essential in developing this synergy. Next, in (Karayel et al., 2022) research address challenges and suggest cutting-edge technology by examining recent advancements in integrating catalytic electrodes for hydrogen production systems. Then, Karayel et al. (2022) investigatedthe cost-effectiveness of producing green hydrogen while identifying important production sites in Kenya. In its continuation, Kazakhstan evaluates capacity to develop a significant green hydrogen generator, concentrating on water use and the significant renewable power requirements (Tleubergenova et al., 2023). Another study suggest plans of Chile to become carbon neutral by 2050 through clean energy programmes and their studies found that off-grid green hydrogen generation is an eco-friendly substitute for traditional techniques(Sharma et al., 2023). Together, the RE sources and hydrogen contributed to the development of sustainable energy with worldwide ramifications.

*(iv). Techno-economic assessment for green energy generation:* In 2022, a researcher examined the possibilities of a RE based coordinated electricity and hydrogen production system with an emphasis on its financial and environmental benefits, particularly in wind farm settings across China (Wang et al., 2022). Another study compared offshore wind power, onshore wind power, and onshore photovoltaic power to determine, which is the most economically feasible approach for producing green hydrogen utilizing various RE sources and electrolysis methods (Shin et al., 2022). A typical 10 MW PEM electrolysis plant with integrated heat recovery and an ORC technology for extra power generation is a significant advancement as suggested in (Vives et al., 2023). A techno-economic analysis of a two-step thermo-chemical water splitting process is conducted in (Lee et al., 2023), with cost-effectiveness, optimization, and technological improvements taken into

account. Examining the green hydrogen generation from gasification of decarbonized biomass with an emphasis on BECCS systems is a novel element (Maciel et al., 2023). In its continued development, Henry et al. (2023)examined the costs of producing wind and H2 while taking into account electrolysis and desalination machinery and providing insights into the effects of various wind conditions.

## MATHEMATICAL MODELLING

In a cement plant, the cement compatible hierarchical substances contains of sand, fly ash, silica fume, and pure tricalcium silicate, etc. At a high temperature above 850°C, the calcination process is an endothermic breakdown reaction of $CaCO_3$ into CaO and $CO_2$. Its chemical reaction could be expressed with (1) (Li et al., 2022),

$$CaCO_3 \rightarrow CaO + CO_2 \Delta H^{\circ} = 178 \text{ kJ/mol} \tag{1}$$

The proposed integrated system of $H_2$ based cement production is assessed depending upon clinker output and amount of $CO_2$ saved by switching from coal to $H_2$ as a fuel. The SEI needed for cement clinker production is calculated using (2) (Li et al., 2022),

$$SEI = \frac{\dot{m}_{H_2} \times LHV_{H_2}}{\dot{m}_{clinker}} \tag{2}$$

where, $\dot{m}_{H_2}$: mass flow rate of $H_2$, $LHV_{H_2}$: low heat value of $H_2$, $\dot{m}_{clinker}$: mass flow rate while clinker cement production.

The power needed for the electric heater, which regulates the water temperature for the electrolyzer at 80°C, is taken from the PV through the power converter. Calculating this power is done using (3) (Nasser & Hassan, 2023),

$$P_{eh} = \dot{m}_{water} \times C_{p\_water} \times (80 - T_a) \tag{3}$$

where, $P_{eh}$: heater power (W); $\dot{m}_{water}$: mass flow rate of water (kg/hr); $C_{p\_water}$: water-specific heat at constant pressure; $T_a$: ambient temperature (°C).

The controller adjusts the necessary power to the electrolyzer and its power is calculated as using (4) (Nasser & Hassan, 2023),

$$P_{ez} = \left(P_{pv} \times \eta_{dc/dc}\right) + \left(P_{bat} \times \eta_{con}\right) - P_{eh} \tag{4}$$

where, $P_{pv}$: output power of PV module; $\eta_{dc/dc}$: dc-dc power conditioning performance efficiency, $\eta_{con}$: power conversion efficiency.

The $CO_2$ emission for each kg of clinker produced is another factor that is used to assess the performance of the integrated system. It can be observed that $CO_2$ emission is a dimensionless parameter and represents the carbon footprint of cement production unit. This is represented by (5) (Li et al., 2022),

$$CO_2\, emission = \frac{\dot{m}_{CO_2}}{\dot{m}_{Clinker}} \tag{5}$$

where, $\dot{m}_{CO_2}$: $CO_2$ emitted mass flow rates; $\dot{m}_{Clinker}$: clinker produced mass flow rates.

The size of diesel generators is often chosen to match the highest load of the demand. To maintain system reliability when the diesel generator is not generating the peak load, it must operate in reduced operating modes. However, this causes the diesel engine to operate inefficiently, consume more fuel, and accumulate more oil debris. This study put up a fresh dispatching approach for diesel generators while taking into account all the aforementioned issues. The generator will operate in an adaptive manner with a steady load and the ideal loading percentage, which can be expressed by (6) (Premadasa et al., 2023),

$$\delta_{fc} = -0.2692 \times f_{LP}^3 + 0.7469 \times f_{LP}^2 - 0.6645 \times f_{LP} + 0.4567 \tag{6}$$

where, $\delta_{fc}$: fuel consumption per unit diesel generator (litre/kWh); $f_{LP}$: optimal loading percentage.

PV panels linked to the DC side of the system produce electricity for the system through the photovoltaic effect. PV panels calculate their hourly energy output depending on the measured temperature and solar radiation. Equation (7) can be used to compute the output of a PV array (Li et al., 2022),

$$P_{pv} = P_{pv\_stc} f_{pv} \left( \frac{G_T}{G_{T\_stc}} \right) \left\{ 1 + \alpha_p (T_c - 25) \right\} \tag{7}$$

where, $P_{pv\_stc}$: PV array related capacity (kW); $f_{pv}$: derating factor of PV system (%); $G_T$: solar radiation incident on the PV array (kW/m2); $G_{T\_stc}$: solar radiation incident on the PV array (kW/m2); $\alpha_p$: power oriented temperature coefficient (%/°C); $T_c$: PV cell temperature (°C).

Battery storage is one of the most quickly evolving energy storage technologies in the world due to its high efficiency, decreased production costs, lack of emissions and decreased environmental impact after installation. In this study, the Li-ion batteries enhance efficiency and a discharge time range. Its charging state could be expressed with equation (8) (Premadasa et al., 2023),

$$SOC_{bs}(t) = \frac{E_{bs}(t)}{E_{bs\_rated}} \tag{8}$$

where, $E_{bs}(t)$: energy capacity at present; $E_{bs\_rated}$: rated energy capacity.

The annual real interest rate is the discount rate used to convert between one-time costs and costs that are spread out over a year. It may be calculated using (9) (Al-Shetwi et al., 2023),

$$\text{Rate of annual real interest } (i) = \frac{i_o - f}{1 - f} \tag{9}$$

where, $i$: real interest rate; $i_o$: nominal interest rate; $f$: annual inflation rate.

The NPC represents the present value of all installation and operating costs incurred over the system's lifespan. The HOMER simulation's optimization results for it can be expressed by (10) - (11) (Al-Shetwi et al., 2023),

$$C_{NPC} = \frac{C_{y\_tot}}{CRF(i,N)} \tag{10}$$

$$CRF(i,N) = \frac{i(1+i)^n}{(1+i)^n - 1} \tag{11}$$

where, $C_{y\_tot}$: system's annual cost ($/year); $CRF(i,N)$: a ratio to find series of annual cash flows; $i$: current interest rate (%); $n$: project life time (year).

The levelized cost of hydrogen ($/kg) should be considered in any financial analysis of hydrogen generation. The following formula (12) can be used to determine the approximate LCOH value (Nasser & Hassan, 2023),

$$LCOH = \frac{CI + \sum_{n=1}^{20} \left\{ \frac{O \,\&\, M + REP}{\left(\frac{i-f}{1+f}\right)^n} \right\}}{\sum_{n=1}^{20} (M)} \tag{12}$$

Where, $CI$: capital investment; $O$ & $M$: operation and maintenance cost; $REP$: replacement cost; $f$: inflation rate and $M$: annual output of hydrogen.

## ENERGY POTENTIAL ASSESSMENT

One of India's fundamental businesses is the cement industry. In 1904, Madras (now Chennai), India began organizing the production of cement. Recently, source materials other than limestone have included seashells, sludge from chemical fertilizers and slag from the iron and steel sector. With more than 7% of the installed capacity worldwide, India ranks second in the production of cement. Domestic cement production increased to 356 million tons in FY 2022 from 296 million tons in FY 2021 (Business Wire, 2023). Cement industry in Chhattisgarh is the backbone of Indian economy. In 2019, it was ranked sixth for ease of doing business. In this state, the Directorate of Industries has 27 locations, CSIDC has 26 locations and Dedicated Industrial Parks has 4 locations to flourish industry. Chhattisgarh has an abundance of key raw resources for cement plants, including gypsum, limestone and clay. In the years (2013-18), 300 lakh tons of limestone were produced on average. The third most abundant mineral produced in the state is limestone (CSIDC, 2019). Most of the cement factory is in Baloda Market and due to this reason Balodabazar district is known as the hub of cement. Its site potential for green energy generation is illustrated in succeeding sections:

In this study, green hydrogen production and application feasibility for a site at Godhi (as shown in Fig. 1) has been evaluated in cement industry. Godhi village is located in Bhatapara tehsil of Raipur district in Chhattisgarh. The location map offers a visual depiction of the region's geography and topography, aiding viewers in understanding the area's physical characteristics and identifying prominent landmarks within the landscape. Following are observations on solar power generation potential at site: It has significant solar energy potential due to its geographical location and favourable climate. The state receives ample sunlight throughout the year, which makes it suitable for solar power generation. It has annual average solar radiation: 5.08kWh/m$^2$/day with annual average temperature: 26.20°C.

*Figure 1. Location mapping of proposed study region among Indian cement industry unit*

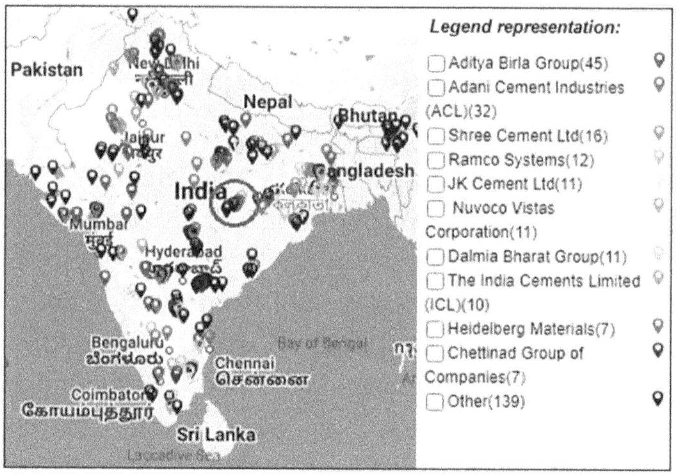

Table 1 and Fig. 2 show the variation of solar GHI and temperature resource at Godhi, Chhattisgarh from January to December. It can be observed that solar GHI resource is maximum in the month of April (i.e., 6.710 kWh/m$^2$/day) and minimum in the month of August (i.e., 3.790 kWh/m$^2$/day) this is because the climate of Chhattisgarh is hot and sun is clear during the month of April and it is rainy season and the sun is not much clear in August. The annual average of the solar GHI resource is 5.080 kWh/m$^2$/day. Similarly, the temperature resource is maximum is 36.04°C in May and minimum that is 18.30°C in December. This is due to the reason that it is summer in May therefore the temperature is high and winter in January therefore the temperature is low. The annual average of the temerature is 26.20°C. The annual average wind resource is 4.38 m/s.

*Table 1. Results illustrating renewable energy source potential at site*

| Month | Clearness index | Daily radiation (kWh/m$^2$/day) | Daily temperature (°C) |
|---|---|---|---|
| Jan | 0.628 | 4.580 | 19.020 |
| Feb | 0.654 | 5.440 | 22.840 |
| Mar | 0.649 | 6.190 | 28.310 |
| Apr | 0.639 | 6.710 | 33.230 |
| May | 0.601 | 6.580 | 36.040 |
| Jun | 0.443 | 4.890 | 31.920 |
| Jul | 0.361 | 3.960 | 27.250 |
| Aug | 0.357 | 3.790 | 26.480 |
| Sep | 0.454 | 4.460 | 26.170 |
| Oct | 0.586 | 5.070 | 24.120 |
| Nov | 0.634 | 4.760 | 20.750 |
| Dec | 0.646 | 4.490 | 18.300 |

*Figure 2. Characteristics of meteorological resources for RE potential at site*

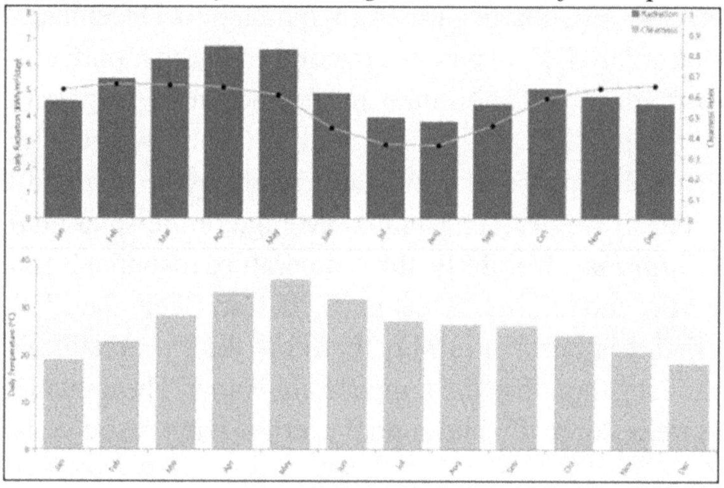

## SYSTEM MODELLING WITH SOLUTION APPROACH

The cement is a powder like substance that may be used with water to aggregate with other materials. The primary constituents are lime, silica, alumina, and iron, which are generated from raw resources including limestone, clay, silica sand and iron sand. The main three steps involved in cement manufacture are the preparation of the raw materials, the production of the clinker and the preparation of the cement itself. Fig. 3 displays a typical basic schematic representation of a cement green energy consumption modeling.

*Figure 3. Flowchart for green hydrogen application analysis for cement plant*

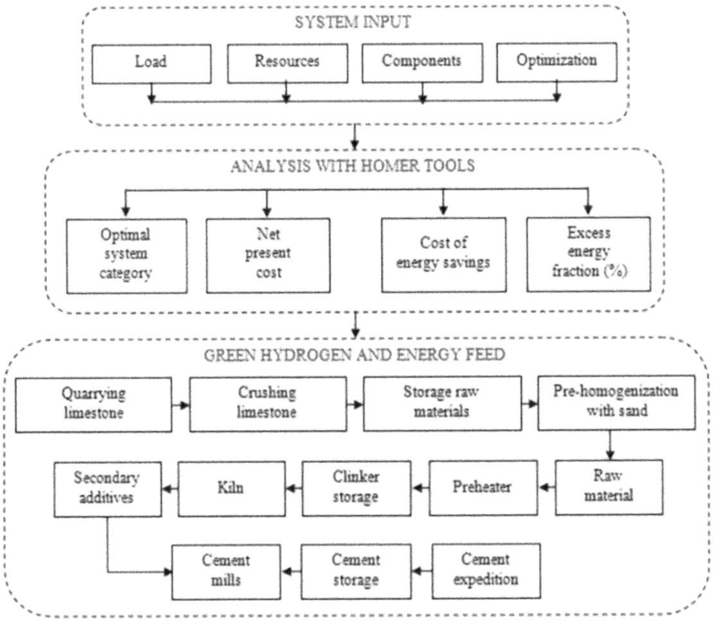


The process of mining and quarrying is known as raw material preparation and it is followed by a mechanical procedure in which crushers and mills are used to reduce the particle size. To achieve the required qualities, each chemical is combined in a predetermined quantity. The raw materials that have been processed are fed into the manufacturing of clinker, which typically entails a preheater, calciner, kiln and cooler. The major and most energy-intensive step in making cement is producing clinker, which is created by turning raw meal through a series of thermodynamic and chemical processes. In order to alter the appropriate qualities, the clinker is next combined with additives and other minerals for final cement manufacture (Juangsa et al., 2022). Fig. 4 presents the block diagram for green hydrogen GEN/PV battery hybrid system. The diagram represents the individual components as well as system integration. All the collectively working in coordination to produce green hydrogen and store it for cement plant demand supply.

*Figure 4. Schematic of the proposed hybrid green hydrogen production model*

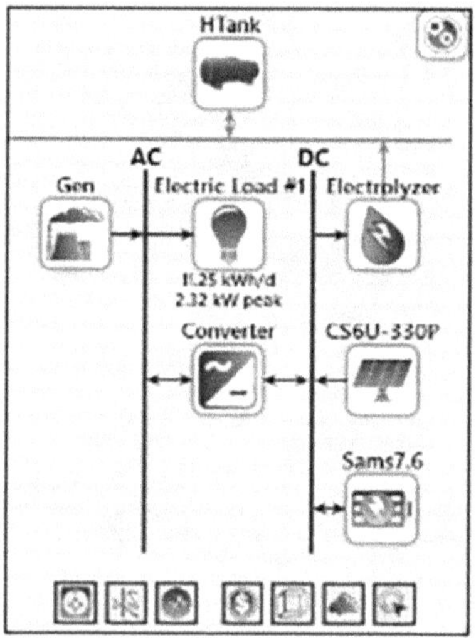


## RESULTS AND DISCUSSION

When developing long-term energy policy, techno-economic modeling is essential. It is critical to have a seamless model for the target region of interest to assure that long-term energy planning and energy goals are consistent. Following are results covering undertaken objectives:

*Figure 5. Characteristics of monthly electricity production and time-power series plot*

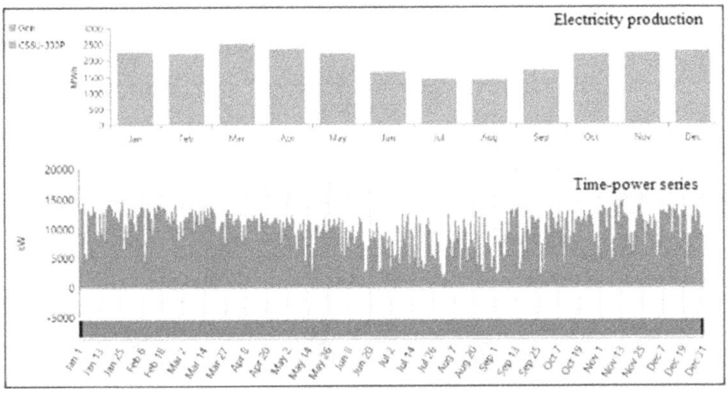


(a). Technical analysis: Table 2 presents the production, consumption behaviour of energy demand for cement plant unit at Godhi. However, Fig. 5 depicts percentage of energy demand fulfilled by various RE sources such as solar, etc. The on-site graph shows the amount of energy produced in MWh for the different months of the year. And Fig. 5 also signifies the time-power plot series of RE sources at site.

*Table 2. Results energy penetration and demand for the studied system*

| Sl. | Features | kWh/yr | % |
|---|---|---|---|
| 1 | Energy production | | |
| | Solar power | 24,234,917 | 20.4 |
| | DG power | 60.5 | 0 |
| 2 | Electricity consumption | | |
| | AC primary load | 4,105 | 20.4 |
| | DC primary load | 0 | 0 |
| | Deferrable load | 0 | 0 |
| 3 | Excess energy | | |
| | Excess electricity | 24,214,428 | 99.9 |

continued on following page

*Table 2. Continued*

| Sl. | Features | kWh/yr | % |
|---|---|---|---|
| | Unmet electric load | 0 | 0 |
| | Capacity shortage | 0 | 0 |
| 4 | Energy penetration | | |
| | Renewable energy fraction and penetration | | |
| | Renewable fraction | 98.5 | |
| | Maximum renewable penetration | 6,852,635 | |
| | Nominal renewable capacity divided by total nominal capacity | - | 100 |
| | Usable renewable capacity divided by total capacity | - | 100 |
| | Total renewable production divided by load | 120,395 | - |

Renewable penetration refers to the percentage of total energy demand that is met by renewable sources such as solar, electrolyzer and generator. Fig. 6 present instantaneous renewable output divided by load, instantaneous renewable output divided by generation, one minus instantaneous non-renewable divided by load.

*Figure 6. Characteristics for electricity demand and supply gap with RE generation*

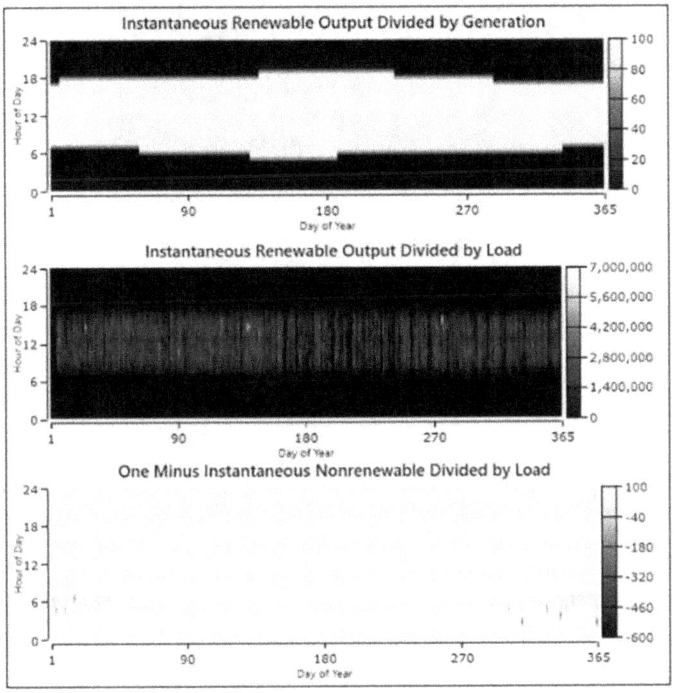


*(a). Solar power generation:* Solar energy works by transforming solar energy into electrical power. Electricity and heat are the two energy types that the sun produces for human use. This study suggests PV system capacity factor is 18.4% with 4,361 hours/year of operation. The total production of solar energy is 24,234,917 kWh/year. Table 4 consist of various parameters essential for PV module performance, however, its output characteristics is shown in Fig. 7.

*Figure 7. PV performance metrics for green hydrogen production*

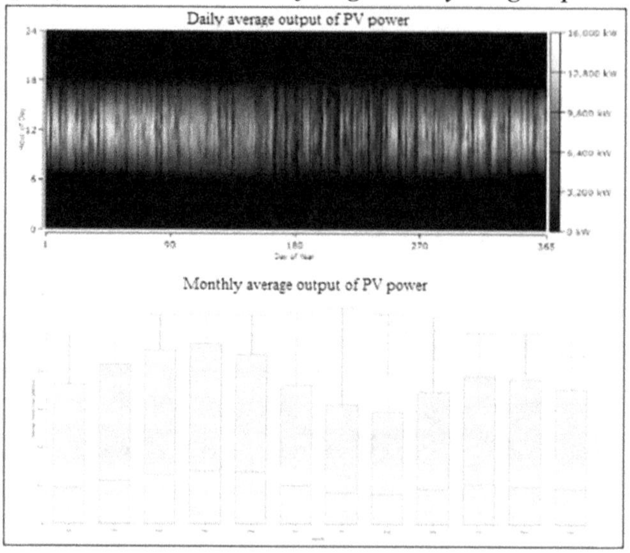


*Table 3. Performance metrics of battery energy storage system*

| Sl. | Features | Value |
|---|---|---|
| 1 | Rated capacity (kW) | 15,000 |
| 2 | Mean output (kW) | 2,767 |
| 3 | Mean output (kWh/day) | 66,397 |
| 4 | Capacity factor (%) | 18.4 |
| 5 | Total production (kWh/year) | 24,234,917 |
| 6 | Minimum output (kW) | 0 |
| 7 | Maximum output (kW) | 14,395 |
| 8 | PV penetration (kW) | 590,336 |
| 9 | Hours of operation (hours/years) | 4,361 |
| 10 | Derating factor (%) | 88 |
| 11 | Levelized cost of energy ($/kWh) | 0.0000120 |
| 12 | Temperature co-efficient | -0.41 |
| 13 | Operating temperature (°C) | 45 |
| 14 | Efficiency (&) | 16.97 |

*(b). Battery storage system:* The batteries in the hybrid RE system are used as energy storage devices to store excess power generated by renewable energy and release the stored power in the event of power shortages. In this work, for battery storage modelling, the capital, replacement and O&M cost are \$500, \$500 and 10 \$/year, respectively. The throughput of the storage system is 45,600 kWh. Its corresponding optimized modelling results and state of charge operation characteristics are given in Table 4 and Fig. 8, respectively.

*Table 4. Performance metrics of battery energy storage system*

| Sl. | Features | Value | Sl. | Features | Value |
|---|---|---|---|---|---|
| 1 | Batteries | 1.00 | 11 | Average energy cost (\$/kWh) | 0 |
| 2 | String size | 1.00 | 12 | Energy in (kWh/yr) | 2,115 |
| 3 | String in parallel | 1.00 | 13 | Energy out (kWh/yr) | 1,910 |
| 4 | Bus voltage (V) | 80.9 | 14 | Storage depletion (kWh/yr) | 6.13 |
| 5 | Nominal capacity (Ah) | 94 | 15 | Losses (kWh/yr) | 212 |
| 6 | Autonomy operation (hour) | 16.2 | 16 | Annual throughput (kWh/yr) | 2,013 |
| 7 | Storage wear cost (\$/kWh) | 0.0116 | 17 | Maximum charge current (A) | 94 |
| 8 | Nominal capacity (kWh) | 7.60 | 18 | Maximum discharge current (A) | 94 |
| 9 | Lifetime throughput (kWh) | 45,600 | 19 | Roundtrip efficiency (%) | 90 |
| 10 | Expected life (year) | 22.7 | 20 | Storage throughput (kWh) | 45,600 |

*Figure 8. Characteristics of battery storage throughout the year for green hydrogen*

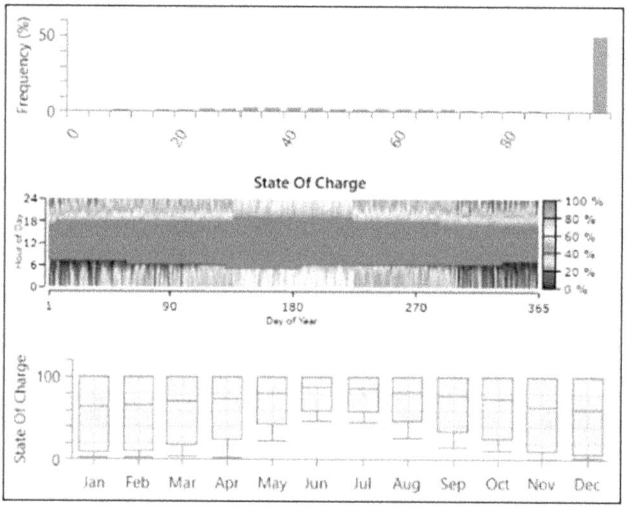


*(c). Power conversion system:* Systems for converting solar energy must incorporate some sort of energy storage to deal with sunlight intermittency. In this cement plant study, the presence of a power converter provides the opportunity to regulate the voltage, frequency and active and reactive power. The capital, replacement and O&M costs of the converter are assumed to be \$300, \$300 and 25 \$/kW/year, respectively. Table 5 and Fig. 9 illustrate the converter performance characteristics for green hydrogen production. It has been assumed that the undertaken converter operate at 95% efficiency with a life span of 15 years.

*Table 5. Specification of power conversion system*

| Sl. | Features | Value | |
|---|---|---|---|
| | | **Inverter** | **Rectifier** |
| 1 | Rated capacity (kW) | 1,000 | 1,000 |
| 2 | Mean power output (kW) | 0.463 | 0.000773 |
| 3 | Minimum power output (kW) | 0 | 0 |
| 4 | Maximum power output (kW) | 2.32 | 0.523 |
| 5 | Capacity factor (%) | 0.0463 | 0.0000770 |
| 6 | Hours of operation (hours/year) | 8,717 | 42.0 |
| 7 | Energy out (kWh/year) | 4,052 | 6.77 |
| 8 | Energy in (kWh/year) | 4,265 | 7.13 |
| 9 | Losses (kWh/year) | 213 | 0.356 |

*Figure 9. Operating characteristics of power converter for green energy production*

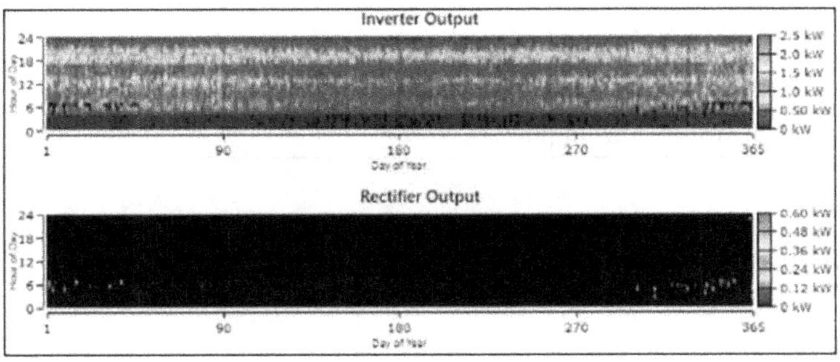


*(d). Hydrogen production for cement plant:* In order to ensure that the hydrogen created by the EL is available for use in the future, it must be stored. This can be done in the FC based system by using an HT. The cost per kilogram of hydrogen capacity could vary depending on the size of the tank. The economics of hydrogen generating facilities has a significant impact on their global expansion. The capital, replacement and O&M costs of the generic electrolyzer are assumed to be \$4226, \$4226 and 0.65 \$/kW/year, respectively. It has a lifetime of 15 years and the efficiency of 80%. India has a huge potential for

solar energy. India's geographical territory surface receives around 5,000 trillion kWh of incident energy annually, with the majority of areas receiving 4-7 kWh/m$^2$ each day. The daily profile of RE availability at site is presented in Fig. 10.

*Table 6. Performance metrics of electrolyzer for hydrogen production*

| Sl. | Features | Value |
|---|---|---|
| 1 | Rated power capacity (kW) | 500 |
| 2 | Mean power input (kW) | 1.83 |
| 3 | Minimum power input (kW) | 0 |
| 4 | Maximum power input (kW) | 500 |
| 5 | Total input energy (kWh/year) | 16,024 |
| 6 | Capacity factor (%) | 0.366 |
| 7 | Mean hydrogen output (kg/hour) | 0.0371 |
| 8 | Minimum hydrogen output (kg/hour) | 0 |
| 9 | Maximum hydrogen output (kg/hour) | 10.1 |
| 10 | Total hydrogen production (kg/hour) | 325 |
| 11 | Specific hydrogen consumption (kWh/kg) | 49.3 |

*Figure 10. Renewable power output daily profile at undertaken site*

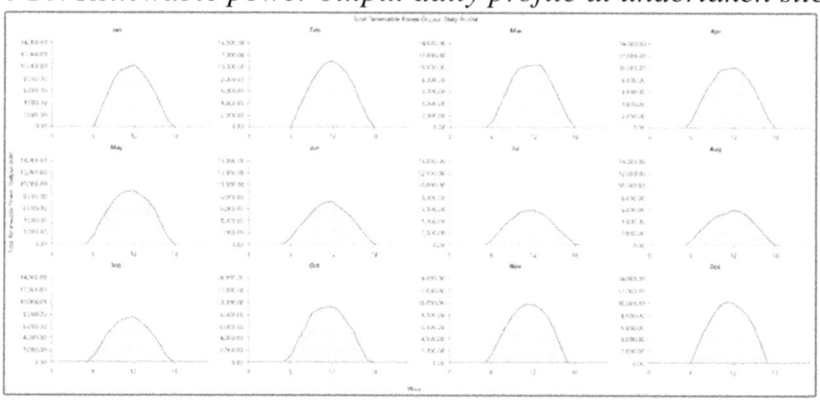

Systems utilizing hydrogen energy look to be among the most efficient options and have the potential to significantly contribute to environmental improvement and sustainability. Additionally, hydrogen has demonstrated potential as a fuel for vehicle using FCs and for a variety of other purposes in the industries. The storage of hydrogen at suitable energy densities presents a technological and financial barrier because of the low density of hydrogen. The capital, replacement and O&M costs of the hydrogen tank are assumed to be $1326, $1326 and $ 68 /kW /year, respectively. The capacity optimization is 500 kg and relative to tank size is 35% for the proposed cement processing unit. Its corresponding results are presented in Table 7 and Fig. 11.

*Table 7. Results for hydrogen production by electrolyzer*

| Sl. | Operation | Value | % |
|---|---|---|---|
| 1 | Electrolyzer production (kg/year) | 325 | 100 |
| 2 | Reformer production (kg/year) | 0 | 0 |
| 3 | Consumption (kg/year) | 325 | 100 |
| 4 | Levelized COH ($/kg) | 4.16 | - |

*Figure 11. Characteristics of electrolyser input and time-power series hydrogen production*

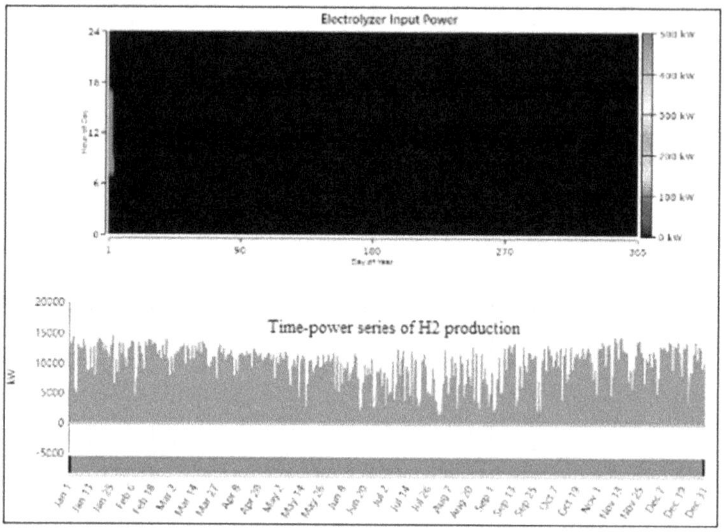


*(ii). Minimization of emission with renewable energy:* An indirect path to cleaner air can be found in the production of green power, which can lower CO2 and other traditional air pollutants by replacing generation from fossil fuels. Energy and related services are increasingly needed to support human social and economic development, welfare and health. In order to meet the energy needs of future generations, the return to RE sources is a great strategy for reducing climate change. The environmental pollutant reduction outcomes are presented in Table 8. Here, citing significance savings with amount of $CO_2$ emission is 72.4 kg/year.

*Table 8. Emission reduction of various gases using hybrid system*

| Sl. | Quantity | Value |
|---|---|---|
| 1 | Carbon dioxide (kg/year) | 72.4 |
| 2 | Carbon monoxide (kg/year) | 0.456 |
| 3 | Unburned hydrocarbons (kg/year) | 0.0199 |
| 4 | Sulfur dioxide (kg/year) | 0.177 |
| 5 | Nitrogen oxides (kg/year) | 0.429 |
| 6 | Particulate matter (kg/year) | 0.00277 |

(iii). Financial analysis: The optimization results provide valuable insights into the performance and efficiency of a renewable energy system, enabling system designers and operators to identify opportunities for improvement and maximize the energy output of the system. Fig. 12 depicts the cost summary of each component integrated to the system, while, Fig. 12 shows the cash flow for the prposed site. It includes replacement, fuel, salvage, operating, capital. In this study, inflation and discount rate have been taken as 8% and 2%, respectively. It suggest expected cash inflows and out flows, replacement cost, salvage value, operating cost, capital over the lifetime of the project, which can help stakeholders evaluate the financial feasibility and profitability of investments in green hydrogen technology.

*Figure 12. Characteristics of optimized component cost and project revenue flow*

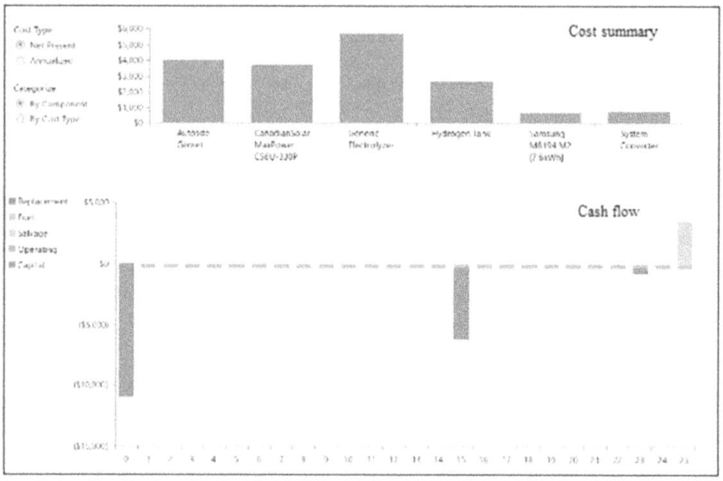


## CONCLUSION

The environment and energy are two of the key issues facing contemporary society. Planning long-term energy policy requires the use of techno-economic modelling. For long-term energy planning and energy goals to be cohesive, the target region of interest must have a correctly and thoroughly built model. This study provided a comprehensive overview of green hydrogen production. Through the utilization of RE sources, such as solar replacing diesel genset unit to power water electrolysis green hydrogen production offers a pollutants emission reduction. In this paper, the cement plant located in Godhi, Chhattisgarh is taken into consideration for evaluation. The study for green hydrogen production of 325 kg/year with renewable sources provides valuable insights and recommendations for cement plant to unlock the transformative power of green hydrogen. In our future energy paradigm, renewable energy is crucial. Thus, by embracing this clean and sustainable energy carrier, industry can pave the way towards a greener and more prosperous future, contributing significantly to the global efforts in mitigating climate change and creating a cleaner planet for generations to come.

## REFERENCES


Abhyankar, N., Mohanty, P., Deorah, S., Karali, N., Paliwal, U., Kersey, J. & Phadke, A. (2023). India's path towards energy independence and a clean future: Harnessing india's renewable edge for cost-effective energy independence by 2047. *2*, 36(5), p.107273.

Al-Shetwi, A. Q., Atawi, I. E., Abuelrub, A., & Hannan, M. A. (2023). Techno-economic assessment and optimal design of hybrid power generation-based renewable energy systems. *Technology in Society*, *75*, 102352. DOI: 10.1016/j.techsoc.2023.102352

Assareh, E., Agarwal, N., Paul, M. C., Ahmadi, P., Ghodrat, M., & Lee, M. (2023). Investigation and development of a novel solar-biomass integrated energy system for clean electricity and liquid hydrogen production. *Thermal Science and Engineering Progress*, *42*, 101925. DOI: 10.1016/j.tsep.2023.101925

Business Wire. (2023). Cement Industry in India 2022 - 2027. https://www.businesswire.com/news/home/20230123005601/en/India-Cement-Industry-Report-2022-Cement-Consumption-Reached-355.46-Million-Tons-in-FY-2022-and-is-Expected-to-Reach-450.78-Million-Tons-by-the-End-of-FY-2027---ResearchAndMarkets.com#:~:text=India%20is%20the%20second%2Dlargest,million%20tons%20in%20FY%202021. Accessed on: January 23, 2023.

Cormos, C. C. (2023). Green hydrogen production from decarbonized biomass gasification: An integrated techno-economic and environmental analysis. *Energy*, *270*, 126926. DOI: 10.1016/j.energy.2023.126926

CSIDC. (2019). Cement Plant.*Chhattisgarh State Industrial Development Corporation*. https://csidc.in/core_sector/aluminum-plant-2/. Accessed on: October 10, 2023.

d'Hubert, X., Henry, F., & Poellaer, S. (2021). Benefits of Micro-grids/Energy Storage for the Cement & Mineral Industries: Part II. The Green Hydrogen Option. In *2021 IEEE-IAS/PCA Cement Industry Conference (IAS/PCA)* (pp. 1-11). IEEE.

Dar, J., & Asif, M. (2023). Environmental feasibility of a gradual shift from fossil fuels to renewable energy in India: Evidence from multiple structural breaks cointegration. *Renewable Energy*, *202*, 589–601. DOI: 10.1016/j.renene.2022.10.131

Dey, S., Sreenivasulu, A., Veerendra, G. T. N., Rao, K. V., & Babu, P. A. (2022). *Renewable energy present status and future potentials in India: An overview*. Innovation and Green Development. DOI: 10.1016/j.igd.2022.100006

Garcia, M., & Oliva, S. (2023). Technical, economic, and CO2 emissions assessment of green hydrogen production from solar/wind energy: The case of Chile. *Energy*, *278*, 127981. DOI: 10.1016/j.energy.2023.127981

Hassan, Q., Sameen, A. Z., Salman, H. M., Jaszczur, M., Al-Hitmi, M., & Alghoul, M. (2023). Energy futures and green hydrogen production: Is Saudi Arabia trend? *Results in Engineering*, *18*, 101165. DOI: 10.1016/j.rineng.2023.101165

He, Y., Guo, S., Dong, P., Lv, D., & Zhou, J. (2023). Feasibility analysis of decarbonizing coal-fired power plants with 100% renewable energy and flexible green hydrogen production. *Energy Conversion and Management*, *290*, 117232. DOI: 10.1016/j.enconman.2023.117232

Henry, A., McStay, D., Rooney, D., Robertson, P., & Foley, A. (2023). Techno-economic analysis to identify the optimal conditions for green hydrogen production. *Energy Conversion and Management*, *291*, 117230. DOI: 10.1016/j.enconman.2023.117230

Herdem, M. S., Mazzeo, D., Matera, N., Baglivo, C., Khan, N., Congedo, P. M., & DeGiorgi, M. G. (2023). A brief overview of solar and wind-based green hydrogen production systems: Trends and standardization. *International Journal of Hydrogen Energy*.

IREA. (2022). Hydrogen from Renewable Power.*International Renewable Energy Agency*. https://www.irena.org/Energy-Transition/Technology/Hydrogen. Accessed on: 16 October 2023.

Ji, W., Zhang, K., Zhan, K., Wang, P., Wang, X., & Yan, Y. (2022). Solar-Powered Environmentally Friendly Hydrogen Production: Advanced Technologies for Sunlight- Electricity-Hydrogen Nexus. *Chinese Journal of Structural Chemistry*, *41*(5), 2205015–2205029.

Juangsa, F. B., Cezeliano, A. S., Darmanto, P. S., & Aziz, M. (2022). Thermodynamic analysis of hydrogen utilization as alternative fuel in cement production. *South African Journal of Chemical Engineering*, *42*, 23–31. DOI: 10.1016/j.sajce.2022.07.003

Karayel, G. K., Javani, N., & Dincer, I. (2022). Hydropower for green hydrogen production in Turkey. *International Journal of Hydrogen Energy*. Advance online publication. DOI: 10.1016/j.ijhydene.2021.10.240

Kumar, S. S., & Lim, H. (2022). An overview of water electrolysis technologies for green hydrogen production. *Energy Reports*, *8*, 13793–1381. DOI: 10.1016/j.egyr.2022.10.127

Lee, S., Na, U. J., & Jo, H. (2023). Techno-economic assessment of green hydrogen production via two-step thermochemical water splitting using microwave. *International Journal of Hydrogen Energy*, *48*(29), 10706–10723. DOI: 10.1016/j.ijhydene.2022.12.119

Li, J., Liu, P., & Li, Z. (2022). Optimal design and techno-economic analysis of a hybrid renewable energy system for off-grid power supply and hydrogen production: A case study of West China. *Chemical Engineering Research & Design*, *177*, 604–614. DOI: 10.1016/j.cherd.2021.11.014

Li, X., & Mulder, M. (2021). Value of power-to-gas as a flexibility option in integrated electricity and hydrogen markets. *Applied Energy*, *304*, 117863. DOI: 10.1016/j.apenergy.2021.117863

Live Mint. (2023). India rolls out green hydrogen production standards. *Live Mint*. https://www.livemint.com/industry/energy/india-rolls-out-green-hydrogen-production-standards-11692426710928.html. Accessed on: 19 Aug 2023

Maciel, L. B. B., Viola, L., de Queiróz Lamas, W., & Silveira, J. L. (2023). Environmental studies of green hydrogen production by electrolytic process: A comparison of the use of electricity from solar PV, wind energy, and hydroelectric plants. *International Journal of Hydrogen Energy*, *48*(93), 36584–36604. DOI: 10.1016/j.ijhydene.2023.05.334

Mazzeo, D., Herdem, M. S., Matera, N., & Wen, J. Z. (2022). Green hydrogen production: Analysis for different single or combined large-scale photovoltaic and wind renewable systems. *Renewable Energy*, *200*, 360–378. DOI: 10.1016/j.renene.2022.09.057

Müller, L. A., Leonard, A., Trotter, P. A., & Hirmer, S. (2023). Green hydrogen production and use in low-and middle-income countries: A least-cost geospatial modelling approach applied to Kenya. *Applied Energy*, *343*, 121219. DOI: 10.1016/j.apenergy.2023.121219

Nasser, M., & Hassan, H. (2023). Egyptian green hydrogen Atlas based on available wind/solar energies: Power, hydrogen production, cost, and CO2 mitigation maps. *International Journal of Hydrogen Energy*.

Osman, A. I., Mehta, N., Elgarahy, A. M., Hefny, M., Al-Hinai, A., Al-Muhtaseb, A. A. H., & Rooney, D. W. (2022). Hydrogen production, storage, utilisation and environmental impacts: A review. *Environmental Chemistry Letters*, *20*(1), 1–36. DOI: 10.1007/s10311-021-01322-8

Pei, Z., Yu, T., Yi, W., & Li, Y. (2021). Twenty-year retrospection on green manufacturing: A bibliometric perspective. *IET Collaborative Intelligent Manufacturing*, *3*(4), 303–323. DOI: 10.1049/cim2.12038

Pocha, C. K. R., Chia, W. Y., Kurniawan, T. A., Khoo, K. S., & Chew, K. W. (2023). Thermochemical conversion of different biomass feedstocks into hydrogen for power plant electricity generation. *Fuel*, *340*, 127472. DOI: 10.1016/j.fuel.2023.127472

Premadasa, P. N. D., Silva, C. M. M. R. S., Chandima, D. P., & Karunadasa, J. P. (2023). A multi-objective optimization model for sizing an off-grid hybrid energy microgrid with optimal dispatching of a diesel generator. *Journal of Energy Storage*, *68*, 107621. DOI: 10.1016/j.est.2023.107621

Ruhnau, O. (2022). How flexible electricity demand stabilizes wind and solar market values: The case of hydrogen electrolyzers. *Applied Energy, 307*, 118194. DOI: 10.1016/j.apenergy.2021.118194

Saedi, I., Mhanna, S., & Mancarella, P. (2021). Integrated electricity and gas system modelling with hydrogen injections and gas composition tracking. *Applied Energy, 303*, 117598. DOI: 10.1016/j.apenergy.2021.117598

Sahoo, N., Kumar, A., & Samsher, . (2023). Design of solar cement plant for supplying thermal energy in cement production. *Journal of Cleaner Production, 426*, 139151. DOI: 10.1016/j.jclepro.2023.139151

Sandaka, B. P., & Kumar, J. (2023). Alternative vehicular fuels for environmental decarbonization: A critical review of challenges in using electricity, hydrogen, and biofuels as a sustainable vehicular fuel. *Chemical Engineering Journal Advances, 14*, 100442. DOI: 10.1016/j.ceja.2022.100442

Sharma, R., Sharda, H., Dutta, A., Dahiya, A., Chaudhary, R., Singh, A., Rathi, K., Kumar, S., Sharma, A., Maken, S., & Nehra, S. P. (2023). Optimizing green hydrogen production: Leveraging load profile simulation and renewable energy integration. *International Journal of Hydrogen Energy, 48*(96), 38015–38026. DOI: 10.1016/j.ijhydene.2023.03.179

Shin, H., Jang, D., Lee, S., Cho, H. S., Kim, K. H., & Kang, S. (2023). Techno-economic evaluation of green hydrogen production with low-temperature water electrolysis technologies directly coupled with renewable power sources. *Energy Conversion and Management, 286*, 117083. DOI: 10.1016/j.enconman.2023.117083

Sousa, V., Bogas, J. A., Real, S., Meireles, I., & Carriço, A. (2023). Recycled cement production energy consumption optimization. *Sustainable Chemistry and Pharmacy, 32*, 101010. DOI: 10.1016/j.scp.2023.101010

Tleubergenova, A., Abuov, Y., Danenova, S., Khoyashov, N., Togay, A., & Lee, W. (2023). Resource assessment for green hydrogen production in Kazakhstan. *International Journal of Hydrogen Energy, 48*(43), 16232–16245. DOI: 10.1016/j.ijhydene.2023.01.113

Torky, M., Dahy, G., & Hassanein, A. E. (2023). GH2_MobileNet: Deep learning approach for predicting green hydrogen production from organic waste mixtures. *Applied Soft Computing*, *138*, 110215. DOI: 10.1016/j.asoc.2023.110215

Van Der Zwaan, B., Lamboo, S., & Dalla Longa, F. (2021). Timmermans' dream: An electricity and hydrogen partnership between Europe and North Africa. *Energy Policy*, *159*, 112613. DOI: 10.1016/j.enpol.2021.112613

Vives, A. M. V., Wang, R., Roy, S., & Smallbone, A. (2023). Techno-economic analysis of large-scale green hydrogen production and storage. *Applied Energy*, *346*, 121333. DOI: 10.1016/j.apenergy.2023.121333

Walter, V., Göransson, L., Taljegard, M., Öberg, S., & Odenberger, M. (2023). Low-cost hydrogen in the future European electricity system–Enabled by flexibility in time and space. *Applied Energy*, *330*, 120315. DOI: 10.1016/j.apenergy.2022.120315

Wang, H., Zhao, X., Zhang, K., & Wang, W. (2022). Economic assessment of a renewable energy-electricity-hydrogen system considering environmental benefits. *Sustainable Production and Consumption*, *33*, 903–920. DOI: 10.1016/j.spc.2022.08.010

## KEY TERMS AND DEFINITIONS

**CCU:** Carbon Capture and Utilization.

**CHP:** Combined Heat and Power.

**DG:** Diesel Generator.

**EL:** Electrlyzer.

**FC:** Fuel Cell.

**FY:** Financial Year.

**GH:** Green House.

**GHI:** Global Horizontal Irradiance.

**HT:** Hydrogen Tank.

**HVO:** Hydrogenated Vegetable Oil.

**LCOE:** Levelized Cost of Energy.

**LCOH:** Levelized Cost of Hydrogen.

**O&M:** Operating and Maintenance.

**ORC:** Organic Rankine Cycle.

**PEM:** Proton Exchange Membrane.

**PtG:** Power-to-Gas.

**PV:** Photovoltaic.

**SDG:** Sustainable Development Goal.

**STC:** Standard Test Conditions.

**RE:** Renewable Energy.

**UNFCCC:** United Nations Framework Convention on Climate Change.

**WT:** Wind Turbine.

**ZEV:** Zero Emission Vehicle.

Chapter 5

# Hydrogen Application of Plasmatron Technology in Fuel Preparation and Mitigation of Exhaust Emissions

**Talip Akbiyik**

*Aksaray University, Turkey*

**Tolga Taner**

https://orcid.org/0000-0002-3065-1942

*Aksaray University, Turkey*

## ABSTRACT

*The tremendous escalation of the global population, the impending inadequacy of current energy resources, and the alarming levels of environmental pollution have heightened the significance of alternative fuels. This condition requires that the alternative fuels under investigation be environmentally sustainable. This chapter indicates hydrogen, which holds a significant position among alternative fuels due to its availability and environmental sustainability. The enhanced combustion velocity and self-ignition temperature diminish the likelihood of knocking in spark ignition engines. The minimal ignition energy facilitates effortless starting. Due*

DOI: 10.4018/979-8-3693-0827-1.ch005

*to the absence of carbon in hydrogen, its exhaust emissions are superior to those of fossil fuel engines. The implementation of plasmatron technology resulted in around 25% fuel savings in vehicles, maintaining vehicle efficiency. The accomplishment was realized through the full combustion of fuel in lean combinations. The complete combustion of fuel diminished vehicle emissions and mitigated environmental harm.*

## 1 INTRODUCTION

A significant contributor to environmental pollution is exhaust emissions from internal combustion engines. The decline in fossil fuel availability, attributed to overconsumption and escalating environmental degradation, has prioritized the investigation of sustainable and renewable alternative fuels. The alternative fuel under investigation must not significantly diminish the performance of the internal combustion engine and should positively impact exhaust emissions. Furthermore, it is essential that this fuel is readily accessible, cost-effective, and compatible with current engines without necessitating significant modifications.

Hydrogen, obtainable from water through electrolysis, possesses superior physical and chemical qualities, offers enhanced engine efficiency compared to gasoline, and has beneficial environmental impacts, rendering it a significant alternative fuel. The utilization of hydrogen as a fuel for engines commenced in the 1920s. Current research indicates that hydrogen can serve as a fuel for conventional automobiles. There are several legitimate explanations for the application's limited adoption thus far. The elevated expense of hydrogen engines and the challenges associated with the secure storage of hydrogen remain unresolved (Bromberg et al., 1999).

Currently, all scientists concur that hydrogen is the optimal energy source capable of sustainably fulfilling the world's escalating energy demands without causing environmental pollution. Hydrogen energy has no risks to human or environmental health. Hydrogen, derived from fossil resources like coal and natural gas, in addition to water and biomass, is seen as both an energy source and an energy carrier (Bromberg et al. 2000). Hydrogen may be generated locally and serves as an energy source applicable across various sectors, including transportation, heating, manufacturing, and culinary applications.

## 2 DEFINITION OF PLASMATRON

Fundamental plasma abnormalities have been examined historically. The efficacy of basic plasmatrons is inadequate, thus they are employed for catalytic procedures in gas composition inside industrial production. It facilitates plasma fuel conversions. It is employed to enhance the transformation of hydrocarbon fuels in the advancement of continuous plasma generation. The rates of reaction have escalated. In plasmatron fuel conversions, the dimensions alter to facilitate successful reformation, hence enhancing the reaction rate and fuel efficiency (Cohn, 2006). Fuels are transformed into hydrogen-rich gases across a broad spectrum. Partial modifications arise from disregarding the generation of plasma soot and are helpful in utilizing rich gases for storage in hydrocarbon fuel. Plasmatron fuel converters encounter issues associated with catalytic reforming (Alharbi et al., 2022). These are the limitations of reaction strength, degradation of fuel components, combustion, soot generation, and a restricted operating temperature range.

The fuel reformer plasmatrons employed in the process facilitate partial oxidation. Under suboptimal conditions, all freed hydrogen atoms react with oxygen and carbon atoms to produce carbon monoxide molecules. The reaction for partial oxidation under sub-ideal stoichiometric circumstances can be given Eq. (1) as follows (Alharbi et al., 2022; Bromberg et al., 2001):

$$CnHm + \frac{n}{2}\,O2 \rightarrow nCO + \frac{m}{2}\,H2 \qquad (1)$$

*Figure 1. Plasmatron (Alharbi et al. 2022)*

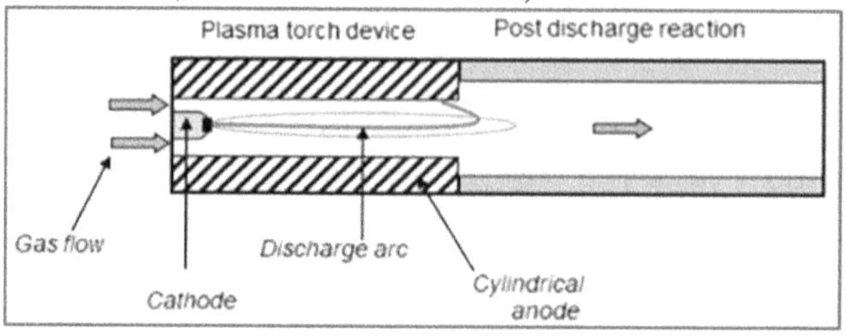


This is due to insufficient oxygen surrounding the carbon atoms in the fuel. The partial oxidation reaction is exothermic in nature (Bromberg et al., 2006). In this instance, around 15% of the calorific content of liquid fuels (gasoline, diesel) is released during partial oxidation reactions. In practice, a greater number of oxygen atoms than carbon atoms are necessary, resulting in an oxygen to carbon ratio (O/C ratio) exceeding one. The process of partial oxidation is exothermic (Alharbi et al., 2022; Bromberg et al., 2006; Cohn 2003).

Plasmatrons can typically be classified into three distinct kinds.

1. High-current thermal plasmatrons
2. Low-current central electrode non-thermal plasmatrons
3. Low-current, wide-area electrode non-thermal plasmatrons

## 3 PLASMATRON TYPES

### 3.1 Thermal Plasmatron

All plasmatrons enhance the temperature of gases by ohmic heating. This temperature partially ionizes the gas. Plasmatrons offer superior control through the electrical heating of gases. As the temperature rises, the plasmatron heat expedites the transformation of hydrocarbon fuels into hydrogen-rich gases. Consequently, ionization and amalgamation are accomplished. Hydrocarbon fuels in hydrogen-rich gases can undergo extensive conversion at elevated temperatures without the necessity of a catalyst. Consequently, it is feasible to eradicate issues associated with catalyst utilization, including limitations in reaction time, poisoning, sensitivity of fuel constituents, and restricted operational temperature (Bromberg et al. 2000).

Enhancing the reaction rate can diminish both the reaction rate and the plasma temperature necessary for fuel conversion. Partial oxidation of steam reforming can occur throughout a broad spectrum of processes. The enhancement of the response rate is accomplished by generating elevated temperature zones exceeding the mean temperature. The elevated temperature of the plasma and the emergence of turbulence inside the plasma significantly enhance hydrocarbon fuel conversions. Plasma reformation might be diminutive because of its elevated density. Swiftly altering plasma characteristics (energy input, flow rate, etc.) are highly effective in meeting the dynamic demands of applications aimed at generating hydrogen-rich gas for automobiles. Immediate creation of hydrogen-rich gas is feasible for application during cold starts (Rajkumar et al., 2024).

*Figure 2. Thermal plasmatron (Bromberg et al., 2002a)*

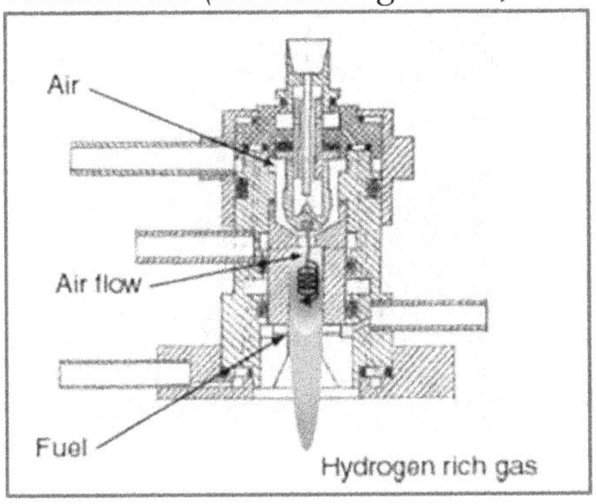


Figure 2 illustrates the configuration of the thermal plasmatron. As air exists as a gas in plasma state within these devices, the operations occur at atmospheric pressure. The processes in plasma operate in direct current mode. The plasmatron comprises a copper cathode, a Hafnium-type anode, and a copper tube anode. The electrodes are divided by electrical insulators. The cathode and anode are cooled by water, resulting in a loss of cooling energy. The measurements indicate that the rise in cooling water temperature is around 70-80% of the plasmatron's output.

Hafnium cathode variants exhibit superior emissivity and extended longevity at currents below 100 A. The plasma current is initiated in the electrode gap. Air is introduced upward into the anode tube, generating a vortex at the plasma electrodes. Alongside the hydrocarbon fuels, air is pumped downward from the electrodes. The heated air mixture and evaporated hydrocarbon fuel are introduced into the plasma reactor.

*Figure 3. DC arc micro-plasmatron fuel-free air operation (Blomberg et al. 2006)*

Figure 3 displays an image of the plasmatron devoid of the reactor cylinder. The injection of fuel and air through the stainless-steel flange during the reformation process is under observation.

*Figure 4. Micro-plasmatron with a reaction extension cylinder and two heat exchangers (Cohn et al., 2006).*

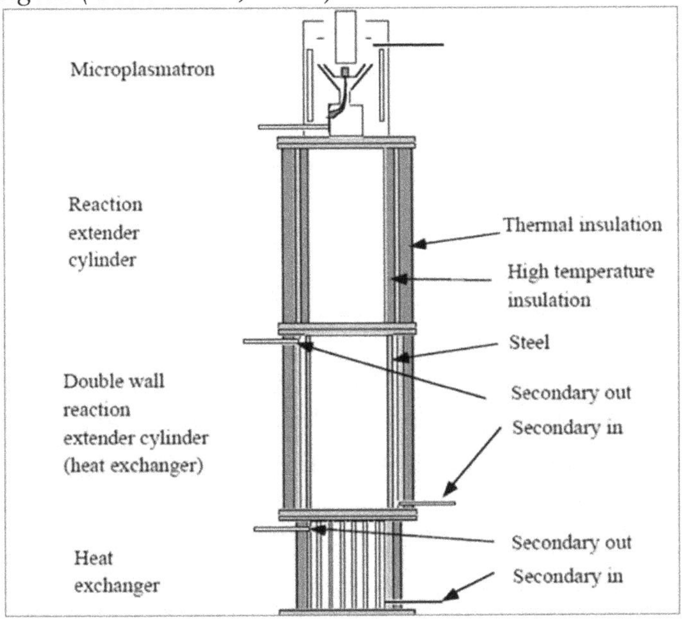


Micro-plasmatron fuel converters comprise a heat exchanger and an elongated cylinder. The heat exchanger can simultaneously utilize air and fuel from the preheater together with rich hydrogen gas. Figure 4 illustrates the micro-plasmatron equipped with a reaction extension cylinder and two heat exchangers.

Preheating the air/fuel diminishes the electrical power demand for the plasmatron and enhances hydrogen generation. Conventional micro-plasmatron fuel converters comprise a steel tube with a diameter of 4 cm and a length of 15 cm, thermally insulated using fiberglass. Table 1 presents the specifications of a conventional plasmatron for direct current arcs. Copper, zirconium, and molybdenum may be utilized to augment plasma production.

*Table 1. DC arc plasma reforming parameters (Blomberg et al., 2006).*

| **Power** | **1.5 – 10 kW** |
|---|---|
| Voltage | 120 – 240 V DC |
| Current | 15 – 75 A DC |
| Air flow rate | 0.5 – 1.5 g/s |
| Fuel | 0.3 – 0.5 g/s |

## 3.2 Low Current Fuel Converter Plasmatrons

Plasmatron fuel converters deliver low current electricity to air and hydrocarbon fuels. The generation of hydrogen-rich gas in hydrocarbon fuel is expedited and augmented. While thermal plasmatron fuel converters enhance enthalpy and accelerate reaction rates, the gas' enthalpy remains unchanged in low current plasmatrons. The application of catalysts is superfluous in low current plasmatrons for the substantial production of hydrogen-rich gas from hydrocarbon fuels. Catalysts are employed to address potential common issues, including restricted working temperature, soot generation, fuel component sensitivity, poisoning, and limited reaction duration (Bromberg et al., 2002a).

Figure 5 illustrates the low current plasmatron. Air and fuel are perpetually introduced into the plasma zone, supplying electricity to be transmitted across the electrode gap. This gadget contains air in a gaseous state in plasma form atmospheric pressure. In direct current operation, the cathode experiences sparking. The anode is a cylinder composed of steel or copper. The electrode lacks water cooling.

*Figure 5. Low current fuel converters (Cohn 2004).*

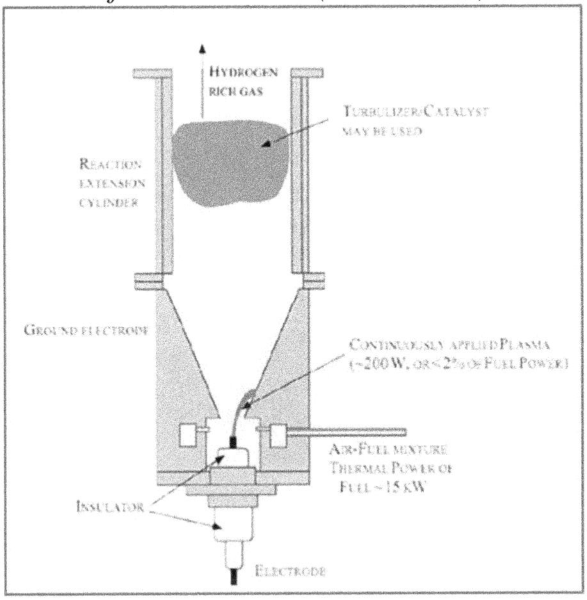


Table 2 presents the plasma characteristics for low-current, low-power plasmatron fuel converters.

*Table 2. Low current concentrated plasmatron fuel converters (Bromberg & Rabinovich, 2001).*

| Define | Quantity and Unit |
|---|---|
| Power | 50 – 300 W |
| Current | 15 - 120 mA |
| $H_2$ flow rate | 30 – 50 l/min |
| Height | 25 cm |
| Volume | 2 l |
| Weight | 3 kg |

The electrical energy required for the condensed plasmatron fuel converter constitutes 2% of the calorific value of the fuel energy input. Low-current devices are compatible with natural gas, gasoline, and diesel fuels. It has been demonstrated that no soot residue persists following the treatment. In standard reforming procedures, the partial enhancement of the process mode achieves an effectiveness of 60-70% (the ratio of the calorific value of the fuel to that of the hydrogen-rich gas).

*Figure 6. Low current plasmatron head (Bromberg et al., 2002b).*

*Figure 7. Low current fuel converter (Bromberg et al., 2002b).*

Diesel fuels can be transformed into hydrogen-rich gases at a power output of 5-20 kW. These conversions can have an efficiency of 80%. Plasma current is essential for the successful conversion of diesel fuel in the plasmatron converter. Nonetheless, plasmatron operation necessitates substantial electrical power and current. The former plasmatrons necessitate reduced power consumption while operation compared to thermal plasmatrons. Figure 6 illustrates the low-current plasmatron head, whereas Figure 7 depicts the low-current plasmatron form.

## 3.3 Low Current Plasmatron Fuel Converters with Large Area Electrodes

This section details low current plasmatron fuel converters featuring large area electrodes. These electrodes function as fuel converters in low-current plasmatrons.

*Figure 8. Low current wide electrode type plasmatron (Cohn et al. 2006).*

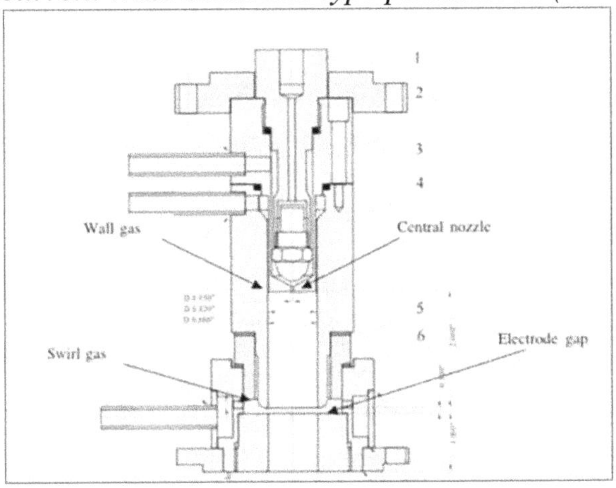


Figure 8 indicates the low current wide electrode type plasmatron in this study. These parts are wall gas, central nozzle, electrode gap and swirl gas.

*Figure 9. Low current wide electrode type plasmatron (Cohn et al. 2006).*

The configuration of the third variant of the plasmatron fuel converter is depicted in Figure 9, with its corresponding photograph presented in Figure 10. The axial gap and the electrode center exhibit symmetry. Gas infiltrates the void, generating a substantial vortex. The objective is to compress the plasmatron to transmit current with this flow. This flow compresses the radial components to direct current towards the device axis and simultaneously rotates the vortex components. This current transmission is generated by alternating current power source with a continuous current. The power supply can generate high voltage and current while maintaining low voltage transmission. In the present phase, the voltage may range from 500 to 2000 V and is subject to temporal variation. The lifespan and duration of the current transmission are extended by elevating the voltage (Bromberg et al., 2006).

In the plasmatron, gas and vortex (plasma) gases are introduced axially through the wall. The vortex gas is introduced along the air walls. The vortex gases are expelled into the fuel, facilitating arc rotation at the electrode tip. The plasmatron possesses several inlets. The direct inlet center is typically utilized for liquid fuels. The liquid within the injector is enveloped by the rapid gas flow. Experiments have demonstrated that optimal performance is achieved when hydrocarbon fuels are combined with vortex gases.

The standard parameters of plasmatron type 3 are presented in Table 3. Type 3 plasmatrons were employed for NOX retention evaluation. ArvinMeritor has developed devices akin to type 3 to improve combustion in spark ignition engines.

*Table 3. Characteristics of low current wide electrode type plasmatron (Bromberg et al., 2003a).*

| **Define** | **Quantity & Unit** |
|---|---|
| Electric power | 200 W |
| O/C | 1.2 |
| Fuel flow rate | 0.8 g/s |
| Corresponding chemical power | W36 |

continued on following page

*Table 3. Continued*

| Define | Quantity & Unit |
|---|---|
| Reformate composition (vol %) | |
| $H_2$ | 7.6 |
| $O_2$ | 1.3 |
| $N_2$ | 64 |
| $CH_4$ | 2.4 |
| CO | 13.0 |
| $CO_2$ | 4.4 |
| $C_2H_4$ | 2.2 |
| $C_2H_2$ | 0.0 |
| $H_2O$ | 7.1 |
| Energy efficiency to hydrogen, CO and light HC | 65% |
| $H_2$ flow rate | 20 l/min |
| Soot (opacity meter) | 0.0% |

## 3.4 ArvinMeritor Plasmatron

The ArvinMeritor Columbus Technology Center produced diesel plasmatron fuel converters following application techniques. The converters achieved a hydrogen concentration of 22% with a catalyst while minimizing soot formation. Plasmatrons are encased in metallic housings that are resistant to little electromagnetic radiation that negatively impacts plasmatron control functions. The power level is entirely regulated by a 250 W power supply in applications that necessitates minimal cooling.

*Figure 10. Gen H plasmatron developed at ARM (Bromberg et al., 2003b).*

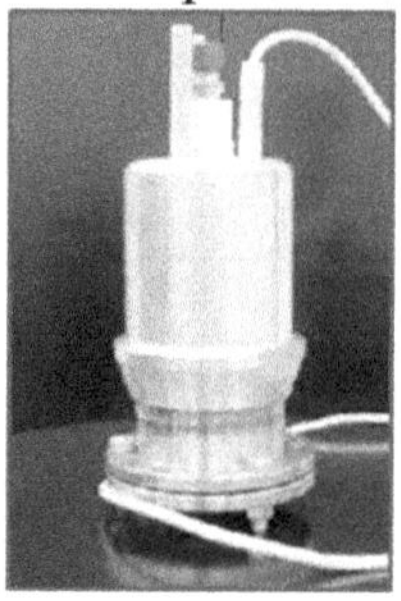

Vehicles are also powered by 12-24 V electrical systems. This plasmatron technology is suitable for cars. Figure 10 depicts the Gen H plasmatron built at ARM.

## 4 CONCLUSION

A variety of factors have propelled the topic of alternative fuels to the forefront. Concerns encompass the rapid global population growth, the imminent depletion of present energy resources, and the severe extent of environmental damage. This suggests that the alternative fuels under investigation are environmentally benign. Hydrogen, the focus of this study, distinguishes itself among alternative fuels due to its environmental benefits and substantial production potential.

Due to its elevated self-ignition temperature and swift combustion rate, spark ignition engines exhibit a little likelihood of knocking. This results from the interplay of factors. The minimal ignition energy facilitates the vehicle's initiation with minimal exertion. Due to the lack of carbon in hydrogen gas, hydrogen engines produce far less pollution compared to their fossil fuel counterparts.

This project employed plasmatron technology to facilitate fuel preparation and reduce emissions. The plasmatron technology enabled automobiles to attain a fuel savings of roughly 25% while maintaining their efficiency. The fuel was combusted in lean mixtures until fully consumed to facilitate this process. Utilizing one hundred percent of the gasoline reduced both environmental effects and vehicle emissions.

## ACKNOWLEDGEMENT

This study is both applied and theoretical in approach, and the authors would like to acknowledge and thank the previously cited scholars for their work on different methods of approaching the application of hydrogen. The authors would also like to thank IGI Global Scientific Publishing for their support.

## REFERENCES

Alharbi, A. A., Alqahtani, N. B., Alkhedhair, A. M., Alabduly, A. J., Almaleki, A. A., Almadih, M. H., Albishi, M. S., & Almayeef, A. A. (2022). Developed plasmatron design to enhance production of hydrogen in synthesis gas produced by a fuel reformer system. *Energies*, *2022*(15), 1071. DOI: 10.3390/en15031071

Bromberg, L., Cohn, D. R., Heywood, J., & Rabinovich, A. (2002a). *On-board plasmatron generation of hydrogen rich gas for diesel aftertreatment and other applications*. Plasma Science and Fusion Center Massachusetts Institute of Technology.

Bromberg L., Cohn D.R., Heywood J. & Rabinovich A. (2002b). Onboard plasmatron generation of hydrogen rich gas for diesel engine exhaust aftertreatment and other applications. Massachusetts Institute of Technology Cambridge MA 02139.

Bromberg, L., Cohn, D. R., Heywood, J., & Rabinovich, A. (2003b). *Hydrogen generation from plasmatron reformers and use for diesel exhaust aftertreatment*. Massachusetts Institute of Technology.

Bromberg, L., Cohn, D. R., & Rabinovich, A. (1999). *Aftertreatment of diesel vehicle emissions using compact plasmatron fuel converter-catalyst systems*. Plasma Science and Fusion Center Massachusetts Institute of Technology.

Bromberg, L., Cohn, D. R., Rabinovich, A., & Alexeev, N. (2000). *Hydrogen manufacturing using low current, non-thermal plasma boosted fuel converters*. Plasma Science and Fusion Center Massachusetts Institute of Technology.

Bromberg, L., Cohn, D. R., Rabinovich, A., Alexeev, N., Samokhin, A., Hadidi, K., Palaia, J., & Margarit-Bel, N. (2006). *Onboard plasmatron hydrogen production for improved vehicles*. Massachusetts Institute of Technology Plasma Science and Fusion Center.

Bromberg, L., Cohn, D. R., Rabinovich, A., & Heywood, J. (2001). Emissions reductions using hydrogen from plasmatron fuel converters. *International Journal of Hydrogen Energy*, *26*(10), 1115–1121. DOI: 10.1016/S0360-3199(01)00049-0

Bromberg, L., Crane, S., Rabinovich, A., Kong, Y., Cohn, D. R., Heywood, J., Alexeev, N., & Samokhin, A. (2003a). *Hydrogen generation from plasmatron reformers: a promising technology for NOx adsorber regeneration and other automotive applications. Diesel Engine Emission Reduction (DEER)*. Workshop.

Bromberg L. & Rabinovich A. (2001). Homogeneous charge compression ignition control by the use of plasmatron fuel converter technology. Massachusetts Institute of Technology Cambridge MA 02139.

Cohn, D. R. (2003). *Cleaner, higher efficiency vehicles using plasmatrons*. Plasma Science and Fusion Center Massachusetts Institute of Technology.

Cohn, D. R. (2004). *Environmental protection and energy savings using plasma technology*. Plasma Science and Fusion Center Massachusetts Institute of Technology.

Cohn, D. R., Bromberg, L., & Hadidi, K. (2006). *Onboard plasmatron hydrogen production for improved vehicles*. Massachusetts Institute of Technology Plasma Science and Fusion Center.

Rajkumar, P., Anand, B. P., Varuvel, E. G., Saravanan, C. G., Josephin, J. S. F., Chinnathambi, A., & Pugazhendhi, A. (2025). Hydrogen-enriched lean-burn strategy for gasoline direct injection engine utilizing non-thermal plasma fuel reformer. *International Journal of Hydrogen Energy, 107*, 512–525. DOI: 10.1016/j.ijhydene.2024.11.070

# Section 3
# Renewable Energy Sources and the Green Carbon Cycle: The Future of Hydrogen

# Chapter 6
# Green Computing for Hydrogen Energy Production as Supplementary Facility in Renewable Energy-Integrated Systems

**Preeti Parihar**
*Department of Electrical and Electronics Engineering, SRM Institute of Science and Technology, Modinagar, India*

**Dharmbir Prasad**
https://orcid.org/0000-0002-9010-9717
*Department of Electrical and Electronics Engineering, SRM Institute of Science and Technology, Modinagar, India*

**Rudra Pratap Singh**
https://orcid.org/0000-0001-7352-855X
*Department of Electrical Engineering, Asansol Engineering College, Asansol, India*

**Ranjeet Singh**
*Department of Electrical Engineering, Faculty of Technology, Delhi University, India*

**Ariba Rizwan**
*Department of Electrical Engineering, Asansol Engineering College, Asansol, India*

DOI: 10.4018/979-8-3693-0827-1.ch006

**Ranadip Roy**
https://orcid.org/0000-0003-2111-2581
*Department of Electrical Engineering, Sanaka Educational Trust's Group of Institutions, Durgapur, India*

**Shubham Verma**
*Software Development Department, Amygb.Ai Pvt. Ltd., Thane, India*

## ABSTRACT

*This research, which provides a comprehensive comparative analysis of green hydrogen generation and storage systems, highlights the importance of renewable energy sources in sustainable energy transition. The outcomes show significant variations in economic viability and environmental effect between the intended Cases-1 and -2. The Case-2, with a greater ROI, IRR and simple payback duration, appears to be a more financially feasible option for using green hydrogen. However, it is important to observe the considerable difference in carbon emissions, with Case-2 generating far more carbon than Case-1. The study's proposed locations are Limali, Gujarat and Bandra West, Mumbai, Maharashtra. Applying renewable energy sources such as Wind/PV/GEN, The ROI is 0.4%, IRR is 0.7% and Simple payback is 24.51 years in the proposed case 1. In the proposed case-2, the ROI is 10.5% and the IRR is 13.3% and the simple payback period is 7.24 years. The carbon emission in Case-1 is 239 kg/year and in Case-2 is 1,438 kg/year. This comparative study gives the best analysis output.*

## INTRODUCTION

Simdega district centrally located in India thrives on agriculture, not merely as an economic pursuit but as an integral way of life for its residents. The agricultural sector holds a pivotal role in supporting the livelihoods of a significant proportion of the local population. Despite its agricultural prominence, Simdega encounters challenges akin to those experienced globally-issues of environmental degradation, resource depletion and the impacts of climate change. The system proposed in this study includes the use of solar as a primary source and grid as the secondary. The grid used in this study purchases as well as sells the energy enduring for a sustainable development in the area. The proposed system does not uses a storage unit as the grid been installed fulfils any need for the use of electricity and also a step towards reducing the capital cost for setting up the system. Jharkhand's Simdega district is home to a wide variety of year-round crops, including grains like rice, maize, and wheat; oilseed crops like Niger, tori, mustard, and groundnut; pulses like black gram, pigeon pea, chickpea and pea; vegetables like lady finger, cowpea, potato, and tomato; and fruits like papaya, mango, guava and jackfruit (Abot, 2020). There are continuous efforts to boost agricultural output through programs like artificial insemination centres for cattle to improve milk yield, despite obstacles including dependency on rain-fed agriculture and poor irrigation infrastructure as shown in Table 1. The temperature and terrain of the district enable a range of crops to be cultivated in certain locations according to the fertility and composition of the soil. In Simdega area, sustainable farming methods are essential for maximizing agricultural productivity and enhancing lives. The gases controlled using the system are carbon dioxide, nitrogen oxide and sulphur dioxide.

*Table 1. Initiatives for agriculture taken in India*

| Sl. | Agricultural segmentation | Sustainable Initiatives | Location | Date | Reference |
|---|---|---|---|---|---|
| 1. | Integrated Portal for Farming Investment | The development of the “Krishi Nivesh Portal” | India | February 29, 2024 | The Hindu Business Line |
| 2. | To promote agricultural exports from India | Value-added millet products. | India | March 6, 2024 | ANI News |
| 3. | Smart agricultural practices | Aims to uplift and empower rural communities. | Jharsuguda, Odisha. | February 28,2024 | Times of India |
| 4. | leveraging technology and market research | Aligning existing agricultural policies with the UN Sustainable Development Goals. | India | October 18, 2021. | The Economic Times |
| 5. | Capsber Agrisciences | Sustainable farming practices for greener future | India | February 29,2024 | The Economic Times |
| 6. | Diagnosis of plant disease severity | Digitizing agriculture with AI, ML and IOT | India | February 5,2022 | Nature |
| 7. | Effective farm methods | Precision Farming SaaS | India | September 23, 2023 | Agriculture post |
| 8. | Modernising and transforming our farming practices | Agri-drone technology | India | December 23, 2023 | Hindustan Times |

continued on following page

*Table 1. Continued*

| Sl. | Agricultural segmentation | Sustainable Initiatives | Location | Date | Reference |
|---|---|---|---|---|---|
| 9. | Region specific smart agricultural technologies | IOT | India | January 21, 2024 | MDPI |
| 10. | Weed removal | Robotics | India | March 20, 2023 | MDPI |

*Paper Layout:* Rest of the paper is organised as follows - energy potential assessment, problem statement, research reviews of green revolution, sustainable agriculture practices, and mathematical formulation, modelling for re powered farming, energy requirements for agricultural, results and finally study concluded.

## Energy Potential Assessment at Site

The Indian state of Jharkhand is home to Simdega, which is extremely important to the area because of its cultural, historical land economic contributions. The customs of several tribal groups, such as the 57,000 person Oraon, Mundaand Kharia, are deeply woven into the district's unique cultural fabric. Each community contributes its own traditions, rituals and creative creations, resulting in a vivid mosaic of Indian cultural expressions that highlights the peaceful coexistence of many ethnic groupings. Simdega's economy is centred on agriculture, which is made possible by the area's rich soil and pleasant environment. Simdega District covers 3761.20 km$^2$ in total as mapping depicted in Fig. 1. In addition to providing for the local population, this agricultural environment is essential to Jharkhand's total food output. The region is proud of its natural beauty and attracts tourists seeking a tranquil vacation. Simdega's biodiversity includes a large range of flora and fauna, highlighting the region's ecological significance and amplifies the allure of its surroundings. Simdega's commitment to social and economic development is exemplified by a number of initiatives, particularly those that improve the lives of indigenous populations. The district's endeavours in the domains of education, healthcare and the establishment of sustainable livelihoods bear witness to its dedication to the overall welfare of its inhabitants. The climate in Simdega is mild and pleasant. In Simdega, it rains very little in the summer and predominantly in the winter. The Koppen-Geiger climate classification places this climate in the Csa category. Simdega experiences 25.1°C temperature on average every year. The average annual rainfall is 1450 mm. December had the least amount of precipitation, averaging 3 mm. With an average of 410 mm, August is when the precipitation peaks. With a temperature of 33.0°C, May is the hottest month of the year. December is the coldest month of the year with a temperature of 17.9°C. Out of the total land 134,024.33 hectares of usable land, 16,367 hectares is irrigated while 117,657.33 hectares is unirrigated.

*Figure 1. Location of Simdega, Jharkhand for agricultural sustainability*

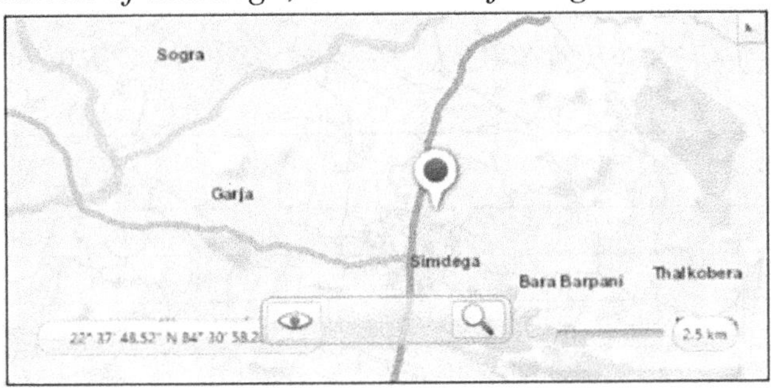


To satisfy its load requirement, Simdega district, like many other places, depends on a combination of energy resources. Geographical features, economic conditions and environmental issues can all affect the availability of energy supplies. Simdega is in the state of Jharkhand, which is well-known for having large coal deposits. Power stations that burn coal may be a substantial source of energy. There may be possibilities for hydropower generation in the area's rivers and other bodies of water. Because of Simdega's unique geographic location and plenty of sunshine, solar energy may be a practical renewable energy source. To capture solar energy, photovoltaic panels are put in place. Wind energy may also be included in the energy mix, depending on regional wind patterns. Electricity may be produced by installing wind turbines. Biomass which includes agricultural leftovers may be burned to produce biogas or utilized in other processes to generate electricity. In order to fulfil Simdega's energy demands, the electrical grid infrastructure's dependability and availability are essential. Energy from several sources may be sent and distributed to end consumers more easily with improved grid connection. Managing the load demand requires both encouraging energy conservation measures and putting into effect energy-efficient activities. This covers programs including those that encourage the use of energy-efficient lighting, appliances and industrial procedures.Location-based mapping involves the utilization of geographical characteristic information for the visualization of data on maps, offering valuable context and insights across diverse domains. The location of Simdega district can be known from the co-ordinates 22.62°N latitude and 84.52°E longitude.

*Table 2. Meteorological resources of study site favourable for weather conditions*

| Month | Clearness index | Daily radiation (kWh/$m^2$/day) | Average wind speed (m/s) | Daily temperature (°C) |
|---|---|---|---|---|
| Jan | 0.629 | 4.460 | 3.950 | 16.640 |
| Feb | 0.653 | 5.320 | 4.310 | 20.370 |
| Mar | 0.628 | 5.930 | 4.720 | 25.810 |
| Apr | 0.622 | 6.510 | 4.390 | 30.790 |
| May | 0.576 | 6.330 | 4.190 | 32.790 |
| Jun | 0.431 | 4.760 | 4.880 | 29.870 |
| Jul | 0.355 | 3.910 | 5.040 | 26.420 |
| Aug | 0.356 | 3.780 | 4.720 | 25.820 |
| Sep | 0.420 | 4.100 | 4.110 | 25.150 |
| Oct | 0.560 | 4.760 | 3.410 | 22.830 |
| Nov | 0.623 | 4.550 | 3.300 | 19.220 |
| Dec | 0.643 | 4.330 | 3.540 | 16.340 |

The charts of solar GHI resource data, which shows the amount of solar radiation received by a surface horizontal to the ground, is important for assessing the solar energy potential of a site and designing the appropriate solar energy system for the site as can be observed from Table 2 and Fig. 2. GHI is a crucial parameter in solar energy applications, especially in the field of photovoltaic and solar thermal systems. It provides information about the amount of solar energy available at a specific location, considering all components of solar radiation. For the clearness index is highest for the month of February with index of 0.653 and lowest month for Julywith 0.355 and the daily radiation data is measured in kWh/$m^2$/ day having peak month for the month of April with a value of 6.510 and the lowest month for the month of August having a value of 3.780

*Figure 2. Meteorological characteristics for the proposed energy system of agriculture supply*

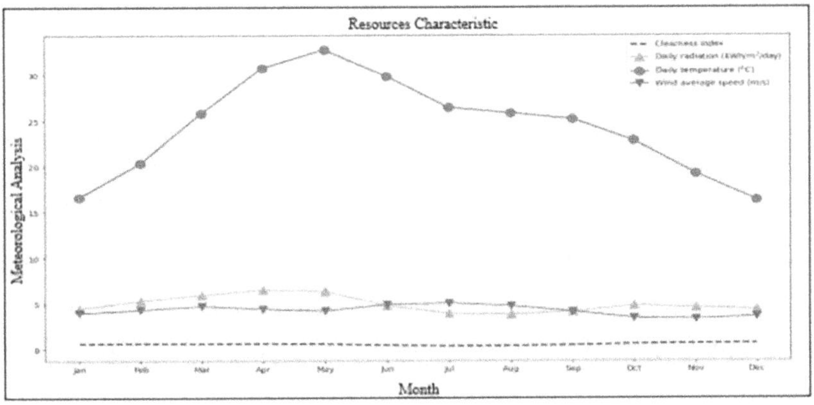


The efficiency and performance of PV solar panels can be strongly impacted by temperature. A number of variables, such as the surrounding environment, solar radiation and solar array design, affect the temperature of PV panels. PV modules feature a temperature coefficient, which expresses how much the module's power output drops as the temperature rises. Usually, the temperature coefficient is given as a percentage (%/°C) of the temperature. When the temperature rises, a positive temperature coefficient means that the power output falls. PV modules' efficiency tends to decline with increasing temperature. The month of May had the highest temperature ever recorded 32.790°C, while the month of December had the lowest with 16.340°C.

The speed record for November is the lowest at 3.300 m/s, while the maximum speed is 5.040 m/s for July. Dust and other particles that might collect on the surface of the PV panels could be carried by the wind under windy and dusty conditions. Due to a decreased capacity to absorb sunlight, this might result in a decrease in the quantity of energy created. Cleaning may be required often in these circumstances. Although wind does not directly contribute to the production of electricity in solar systems, wind speed can have an indirect impact on the total amount of energy generated in these systems. The average wind speed in the area is 4.213 m/s.

## Problem Statement

A problem statement for sustainable agricultural development identifies a specific issue or difficulty that impedes environmentally friendly farming techniques or food production systems. It is simply defining the problem's background and relevance, laying the groundwork for focused solutions to improve the sustainability of agricultural operations.

- *System technical compatibility:* The economic, environmental, and social advantages of using solar energy in the area are demonstrated by the successful completion of solar-powered projects for irrigation, electrification, and the production of green electricity in the Simdega district.
- *Economic feasibility evaluation:* A concept, project, or solution's economic feasibility is evaluated by considering a number of factors, including operational, legal, financial and technological considerations. This assessment is essential to ascertain whether a plan is feasible given the state of technology, the economy, the time period, and the demands and preferences of society.
- *Environmental impact assessment:* To guarantee the long-term viability and sustainability of solar efforts in Simdega district and beyond, it is imperative to tackle issues including community acceptability, operating methods, and completing thorough assessment.

## Research Reviews of Green Revolution

For farmers who might not be able to purchase conventional diesel pumps, these devices offer sustainable water solutions by effectively irrigating vast tracts of land. Following are the research's related to the study:

*(a). Green Energy Growth:* Through wise energy management, economical usage, and renewable technologies like solar, wind, and water, the review research highlights greener power (Bhuiyan, 2022).

While tackling challenges like CEP and FFC, the research places a strong emphasis on ICTs and sustainable NR management for green growth in KBEs (Wang et al., 2023).

*(b). Digital Transformation:* Investigating the relationship between Industry 4.0 and sustainability, the study highlights the significance of sustainability in intelligent manufacturing as well as their tight relationship (Furstenau et al., 2020). The study examines how employee performance on digital strategy goals is impacted by IEO, indicating that risk-taking and proactiveness have favourable effects while creativity has negative effects (Ritala et al., 2021). The research presents novel examples of waste management digitalization in Indonesia and other countries, showing how old approaches are being replaced by the CE using digital platforms (Kurniawan et al., 2022). The study examines the REMs industry in China and discovers beneficial connections between DT, SD, and ES. For competitiveness and sustainability, ES and DT integration is essential (Guan et al., 2023). The study does an extensive literature assessment on industrial digitalization, including suggestions for future lines of inquiry and managerial and scholarly insights (Matt et al., 2023). Greenhouse gas emissions and cost savings from digitalization support the objectives of the circular economy (Kurniawan et al., 2023). ES, DT, and SD were shown to have substantial positive correlations in the PLS-SEM study that examined ES's influence on DT and SD in China's REMs industry (Guan et al., 2023). According to the study, there is a general expectation that digital technology would drastically change cities, infrastructure, buildings, architecture, and construction. It also maps out many images of digital construction (Braun & Kropp, 2023). By mentoring staff members, supporting learning, and advocating for the circular economy, lean leadership nurtures a culture of continuous improvement with an emphasis on coaching, empowerment, and respect (Gatell & Avella, 2024). The study examines how natural resource usage and e-commerce affect green growth in fifteen developing Asian nations, highlighting the positive effects of foreign investment and e-commerce as well as the negative implications of urbanization and mineral consumption (Sun et al., 2024). This research explored the intricate relationship between media, cultural industries and their

impact on mineral resource efficiency through an extensive analysis spanning from 2000 to 2020 across ten major global CO2 emitters (Liu et al., 2024). Ecological sustainability and mitigating climate change are given special attention in this research, which examines China's natural resources, technological innovation, and economic growth (Zhang & Xie, 2024).

*(c). Circular Economy:* In order to demonstrate way to BDA meets the goals of the circular economy by boosting competitiveness and societal value, the study looks at five case studies in emerging nations (Modgil et al., 2021). In order to bring waste management and the circular economy into line with the UN SDGs, the study looks at how COVID-19 affects the SDGs (Sharma et al., 2021). The study aspired to develop a theoretical model linking key resources for I4.0 adoption, sustainable production and circular economy (Bag et al., 2021). The present analysis reveals that research in this domain is fragmented across interdisciplinary fields (Chauhan et al., 2022). In line with responsible consumption, reducing environmental effect, and Vietnamese low-carbon goals, the research focuses on SMEs using CE for sustainable development (Chowdhury et al., 2022). Real technical and economic change is needed to achieve carbon neutrality. Through the decoupling of growth from resource depletion, the CE promotes regenerative economic cycles (Kurniawan et al., 2022). The study investigates the relationship between AM and the CE (Hettiarachchi et al., 2022). Considering the social and environmental crisis being faced and since we know that MNEs are highly pressured toward financial performance researchers, managers and investors should be predominantly oriented to an agenda that encompasses social and environmental sustainability and how they can also align with financial enhancement of the firm (de Oliveira et al., 2023). Implementing circular economy initiatives in business activities is essential for improving the organization's efficiency through resource management as well as protecting the environment (James et al., 2023). The paper discusses Industry 4.0 social challenges and offers solutions. Job loss, a lack of skilled workers, a lack of IT know-how, managerial support, cyber security, and goal alignment are major obstacles (Upadhyay et al., 2023). Intensive collaborative

research accentuates the significance of devising solutions tailor-made for MSW concerns (Munir et al., 2023).

*(d). Conservation of Agriculture:* The authors used a proven research method to collect published articles from reputable databases such as WoS, Scopus and Google Scholar. The authors sought to answer three research questions using SMEs (Truant et al., 2024). The discipline of assessing impacts of changes in farm practice using the 10 indicator structure (later 11) has been helpful in that it permitted science-based, quantitative and qualitative environmental, social, economic and commercial assessments to be made within a single structure (Pretty et al., 2008). The Australian Bureau of Agricultural and Resource Economics have defined similar sustainability indicators of farm performance using a similar methodology and annual farm survey data as employed here (ABARE, 1999) (Dillon et al., 2010). The CA-based cropping system and holistic system approach not only conserves natural resources but also helps in producing more at low costs, improves soil health, promotes timely planting and ensures crop diversification, reduces environmental pollution and adverse effects of climate change on cereal systems in IGP (Jat et al., 2021). The concept of sustainable agriculture is predicated on a delicate balance of maximizing crop productivity and maintaining economic stability, while minimizing the utilization of finite natural resources and detrimental environmental impacts (Chel & Kaushik, 2011). The idea is to identify a set of thresholds that any agricultural production strategy must meet beyond which unsustainable trends caused by the farming systems and associated rage rating of more than 1 for farmer satisfaction and resource conservation is required for a farm to be sustainable technologies would lead to tipping-point phenomena (Koohafkan et al., 2012). The first concerned practitioners' different understandings of modernization. Indeed, the practitioners on the ground showed extremely heterogeneous understandings of modernization (Knickel et al., 2017).

## Sustainable Agriculture Practices

There are sixteen sustainable farming techniques that have been defined by the Council on Energy, Environment and Water. Although acceptance varies, there has been improvement in areas such as organic farming in Sikkim and zero-budget natural farming in Andhra Pradesh (The Times of India, 2023). The government encourages natural farming, which uses age-old, chemical-free techniques, to address these problems. This encourages a variety of bacterial communities in the soil, which improves plant development, resistance to environmental stressors and natural insect resistance (The Wire, 2023). There will be pressure on agriculture to meet future targets, so innovation will be important.Participating in bio fuel will provide farmers with an opportunity to improve economic efficiency. It will also create competition between food and fuel producers for valuable crops (S&P Global, 2023). Economic inequities grew as a result of concentration in some areas, which marginalized small farmers. The revolution accidentally increased wealth disparity in the dairy industry while attempting to increase rural incomes and food supply (Mint, 2023). In contrast, Mizoram, Kerala and Andhra Pradesh, Madhya Pradesh and West Bengal excel in sustainability thanks to better infrastructure and agricultural diversification (The News India Express, 2023).

India, a leader in the Green Revolution, now has to modernize agriculture in a way that is sustainable. In order to maintain environmental sustainability in the twenty-first century, this essay examines the possible advantages and difficulties of incorporating electric tractors into Indian farming techniques (The Sunday Guardian, 2023). Two of the most important issues facing the 21st century are climate change and food poverty. Heat waves and droughts are examples of climate-related occurrences that have an influence on agriculture and increase food demand due to population expansion and dietary changes. Farmers adjust to reduce climate risk, but a comprehensive strategy is required. In light of climatic issues, agricultural productivity has to rise by 60% by 2050 to fulfil food demand (The Hindu, 2023). The Indian agriculture sector has been inching close to achieving these figures on the back of the green revolution and tech-backed agriculture solutions that are integrally playing a key role in making sustainable agriculture a reality (The Financial Express, 2023).

This vital economic sector is also part of the problem as it contributes 17% to greenhouse gas emissions, produced especially by intensive farming characterized by efficient processes that maximize production yield (Vatican News, 2023).

## Mathematical Formulation

Understanding and assessing the risks associated with farming requires the use of mathematical models, which help farmers use resources optimally to increase produce while lowering expenses and environmental effect. When exposed to sunlight, arrays of panels continually generate power. Three critical parameters affect the design of a PV cell: the idealist factor, short-circuit current and open-circuit voltage. The output current produced by the PV panel is calculated by equation (1) (Bharathi & Sasikumar, 2021),

$$I_0 = I_{irr} - I_{01}\left(e^{\frac{(V_0+IR_1)}{\left(N_a \times \frac{kT}{q}\right)}}\right)^{-1} - \left(\frac{V_0 + IR_1}{R_2}\right) \quad (1)$$

where, $I_0$: output current, $I_{01}$: reverse diode current, $V_0$: output voltage, $N$: PV cell number, $a$: ideality factor, $\frac{kT}{q}$: Boltzmann constant.

The inverter synchronizes the AC output with the local electrical grid's frequency and phase in addition to converting DC electricity to AC voltage. Furthermore, a MPPT technology that optimizes the energy harvest from PV panels by adjusting voltage and current levels to the panels' maximum power point, is frequently included in inverters. The converter's efficiency is represented by equation (2) (Hassan et al., 2023),

$$\eta_{ec} = \frac{P_{opc}}{P_{ipc}} \quad (2)$$

where, $\eta_{ec}$: converter efficiency, $P_{opc}$: output power of the converter, $P_{ipc}$: input power of the converter.

The goal of reducing agriculture's dependency on energy is to increase energy efficiency in the use of machinery, fertilizer and land management practices. But because of the rebound effect, these developments can paradoxically lead to a rise in energy usage. This happens when manufacturers choose more energy-intensive processes as a result of efficiency advancements, which raises the amount of energy used. Agriculture energy efficiency is represented by the equation (3) (Han & Zhou, 2024),

$$EF_{it} = \frac{OEI_{it}}{AEI_{it}} \tag{3}$$

where, $EF_{it}$: agriculture energy efficiency around region $i$ in $t$ year, $OEI_{it}$: agricultural energy input at the optimal production frontier around region $i$ in year $t$, $AEI_{it}$: actual agricultural energy input in $i$ region for year $t$ .

The NPC of a system is the total present value, discounted to present value, of all costs incurred during the system's lifetime for its acquisition, implementation, operation, and maintenance, less any savings or benefits realized during that same period. This system is given by the equation (4) (Pal & Mukherjee, 2021),

$$NPC = \frac{TC_{an.\cos}}{C_{RF}\left(i, R_{project}\right)} \tag{4}$$

where, $NPC$: net present cost, $TC_{ancos}$: total annualized cost ($/year), $C_{RF}$: capital recovery factor.

The levelized COE is given by the equation (5) (Pal & Mukherjee, 2021),

$$LCOE = \frac{TC_{ancos}}{E_{ser}} \tag{5}$$

where, $LCOE$: levelized cost of energy, $TC_{ancos}$: total annualized cost ($/year), $E_{ser}$: total electrical load served per year (kWh/year).

## Modelling for RE Powered Farming

Farmers may increase their resistance to agricultural hazards, increase profitability, and use resources more efficiently by utilizing mathematical models to guide their decision-making.

*Current System:* JGF5+F75, Simdega, Jharkhand 835223, India's power demands are satisfied. The annual cost of energy for operation is presently \$6,039.

*Proposed System:* We intend to add 680 kW of PV, this would cut the running costs to -\$103,172 per year. The investment has a payback period of 4.58 years and an IRR of 21.7%.

Block diagrams provide an emphasis on learning and simplicity. The system modelling is illustrated in Fig. 3.

*Figure 3. Block diagram using Grid/Solar PV/Converter and electric load for agriculture supply*

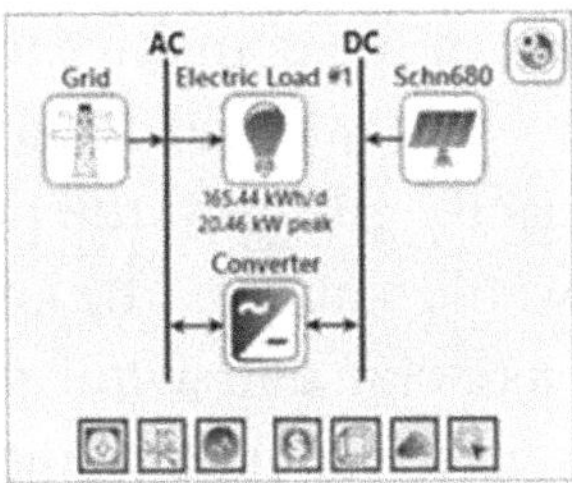


This components description illustrate agriculatural power supply. It entails outlining each component's features, capabilities and interactions with other parts within the overall structure:

*(a). PV Module:* Photovoltaic power generation is the process of using photovoltaic cells also commonly called solar cells to convert sunlight into electricity. The PV model used here is Schneider ConextCoreXC 680kW with Generic PV with a rated capacity of 680.08 kW and comes under flat plate category. The capital cost for the unit is 500,000\$ and the cost for the O&M is 10\$. The temperature coefficient for

the PV system is 0.4100 while it has operating temperature of 45°C. The system has derating factor of 96% and an efficiency of 17.30%.

*(b). Power Converter:* In a photovoltaic system, an inverter's main job is to change the direct current produced by the solar panels into alternating current. AC is used by most electrical equipment and the power grid. The inverter synchronizes the generated AC with the utility grid in PV systems that are connected to the grid. This enables extra energy to be supplied back into the grid and through net metering, the system owner may get paid or earn credits. PV system's total efficiency is enhanced by inverters. Superior inverters enhance the system's overall performance by converting DC to AC with the least amount of energy loss. Inverters can occasionally be found in hybrid systems that combine solar energy with backup power sources like batteries or generators. The energy transfer between these various parts is controlled by the inverter. The capacity of converter used here is 50,000 kW as to transmit maximum energy to the grid without the losses because there is no battery used to store the energy. The capital and replacement of the converter used is $300.

*(c). Grid Connectivity:* A growing alternative is the distributed grid, which generates electricity closer to the point of consumption, frequently utilizing renewable sources. The issues posed by intermittent energy sources such as solar and wind power must be addressed in order to integrate renewable energy into the system. In order to manage these difficulties, energy storage systems and advanced grid technology are important. Digital technology is used by smart grids to improve reliability, efficiency and performance. The grid power price is 0.100 $/kWh and the grid net excess price is 0.050 $/kWh. Better control of the supply and demand for energy is made possible by their ability to provide two-way communication between utilities and customers. To provide a steady and dependable electricity supply, grid stability is essential. Demand variations, abrupt changes in generation and fault-related disturbances are a few examples of the variables that might affect stability.

## Energy Requirements for Agricultural

The goal of sustainable agriculture is to meet society's current food and textile demands while preserving the capacity of present or future generations to meet their own needs. Being a large power user, agriculture has a number of issues that impact the availability of resources. Observation for the present study is summarised as in the succeeding sections:

*(i). Energy Consumption:* Every piece of machinery, input, and end product used in agriculture has energy. In order to bring agriculture closer to sustainability, energy analysis can be used to optimize energy consumption and boost energy efficiency. It also makes it possible to quantify the amounts of energy required for agricultural output. In this analysis, Table 3-4 present electricity produced and it shows that the PV produces maximum electricity which is 1,160,751 kWh/year and it produces 97.7% of the total electricity while the electricity purchased from the grid is around 27,592 kWh/year which contributes to only 2.32% of the total electricity which is 1,188,343 kWh/year. Fig. 4 shows load profile under various circumstances. For energy systems, an electrical summary could provide a snapshot of essential details about the electrical infrastructure, power generation methods, distribution networks and patterns of energy consumption.

*Table 3. Energy production summary at the proposed energy system of agriculture load supply*

| Production | kWh/year | % |
|---|---|---|
| Schneider ConextCoreXC 680kW with Generic PV | 1,160,751 | 97.7 |
| Grid purchases | 27,592 | 2.32 |
| Total | 1,188,343 | 100 |

*Figure 4. Load characteristics under various agricultural conditions*

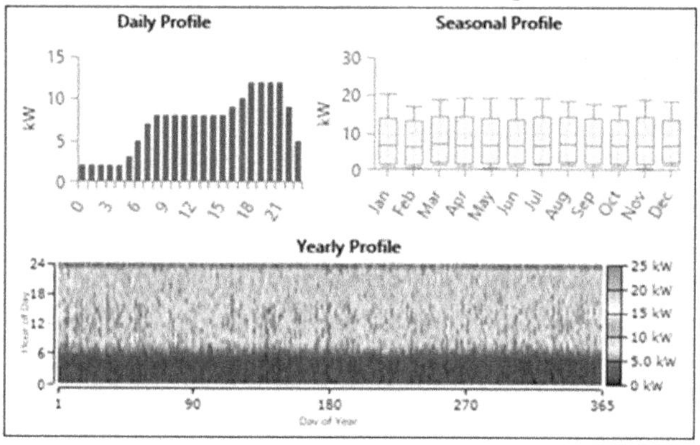


*Table 4. Energy consumption results for proposed energy system of agriculture load supply*

| Consumption | kWh/year (%) | Consumption | kWh/year (%) |
|---|---|---|---|
| AC primary load | 60,386 (5.34%) | Excess Electricity | 183 (0.0154%) |
| DC primary load | 0 (0%) | Unmet Electric Load | 0(0%) |
| Deferrable load | 0 (0%) | Capacity Storage | 0 (0%) |
| Grid sales | 1,069,919 (94.7%) | Renewable Fraction | 97.6(%) |
| Total | 1,130,305 (100%) | Max. Renewable Penetration | 105 (%) |

*(ii). Power Demand Estimation:* The goal of this study is to model and forecast the energy consumption in one of the village of Indian agriculture, which is a crucial component of production. Following are results compilation given below:

(a). Electrical Summary: Photovoltaic systems offer a renewable and sustainable source of electricity generation, harnessing the abundant energy of the sun to produce clean electricity with minimal environmental impact. Fig. 5 shows the electricity produced from the renewable source used. The microgrid requires 3101 kWh/day and has a peak of 646 kW. In the proposed system, it shows the generation sources serving the electrical load.

*Figure 5. Characteristics for energy consumption and production*

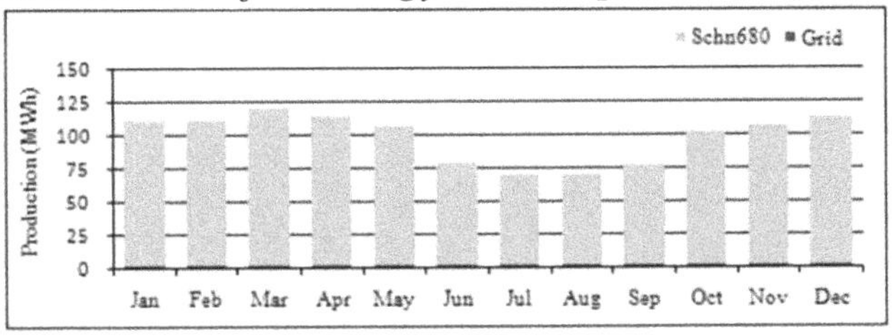


From Table 5, it can be clearly analysed that the renewable fraction has 97.6% value and the maximum renewable penetration is 105% because it represents the extent to which the solar renewable energy contributes to the overall energy mix in the area for Schneider ConextCoreXC 680kW with Generic PV.

*Table 5. Renewable penetration for energy production*

| Quantity | Value |
|---|---|
| Renewable fraction (%) | 97.6 |
| Maximum renewable penetration (%) | 105 |

(b). Variation of Power Demand and Generation: The characteristics acts as an aid in this researchin a wide range of scientific and technical domains by providing essential insights into how systems, signals, and variables interact over time as presented in Fig. 6.

*Figure 6. Characteristics of the combined heat and power for agriculture supply*

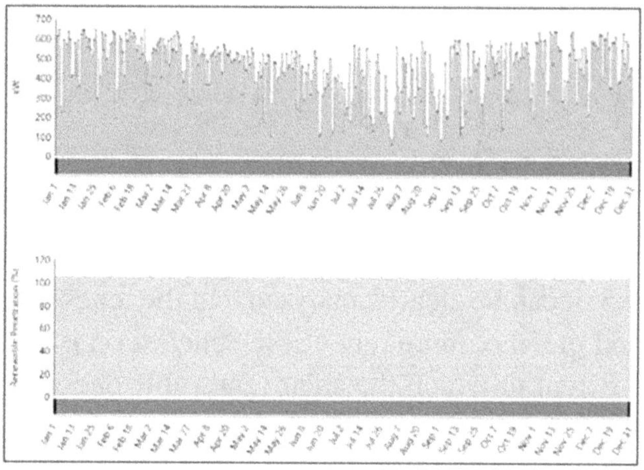

The objective of this approach is to improve sustainability in the agriculture industry by decreasing the dependence on fossil fuels, minimizing greenhouse gas emissions and maybe lowering the operating expenses for farmers as presented in Fig. 7.

*Figure 7. Characteristics of the renewable resource penetration for agriculture supply*

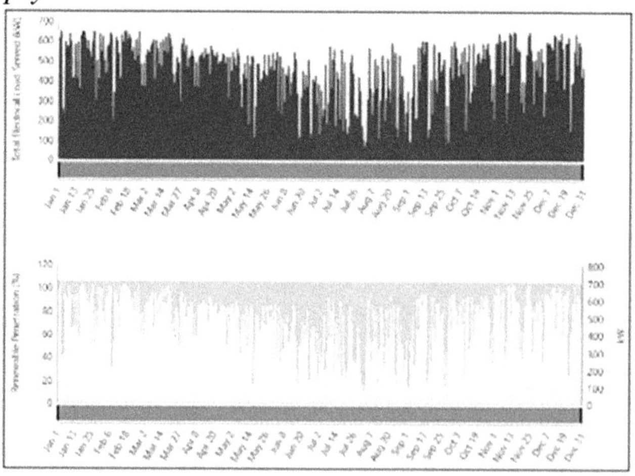

## RESULTS AND DISCUSSION

The debate surrounding the climate, water, land, energy and food nexus revolves around the agricultural sector. The primary goal in creating a sustainable agriculture sector is to feed the growing global population while minimizing its negative effects on the environment and protecting the most valuable natural resources for coming generations. Following are the outcome described below:

*(i). Technical Feasibility Analysis:* Technical compatibility is essential to understand efficient system integration. Following are various technical aspects:
*(a). Solar Power Generation:* The power output of PV system is the amount of electrical power it generates under specific conditions. The power output is influenced by various factors and understanding these factors is crucial for designing, installing and optimizing the performance of a PV system as given in Table 6-7. The total production of electricity is 1,160,751 kWh/year and the mean output is 133 kW. The PV penetration is 1,922% and the hours of operation are 4,372 hrs/year.

*Table 6. Electricity production for running various equipment's in agricultural work*

| Quantity | Value |
|---|---|
| Rated Capacity (kW) | 680 |
| Mean Output (kW) | 133 |
| Mean Output (kWh/d) | 3,180 |
| Capacity Factor (%) | 19.5 |
| Total Production (kWh/year) | 1,160,751 |
| Dedicated Converter (kW) | 680 |

*Table 7. Power output chart for proper utilization in the agricultural fields*

| Quantity | Value |
|---|---|
| Minimum output (kW) | 0 |
| Maximum output (kW) | 680 |
| PV Penetration (%) | 1,922 |
| Hours of operation (hrs/year) | 4,372 |
| Levelized Cost ($/kWh) | 0 |

*Figure 8. Characteristics for solar power generation and power converter performance*

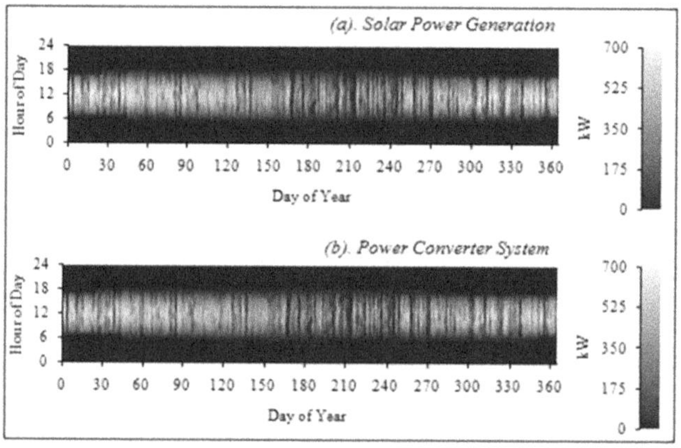


A photovoltaic system's daily power production may be estimated by considering factors including a gradual rise in power at sunrise, a peak output at solar noon and a decrease in power at nightfall due to varying levels of irradiance. Fig. 8 shows the power output characteristics of the PV with power converter performance. Actual production is influenced by panel efficiency, system architecture and site conditions. For precise estimations, use solar modelling software that includes the precise location and panel parameters.

*(b). Power Converter Simulation Output:* In a renewable energy system, power converters provide energy produced by solar or wind turbines into a form that may be used locally or connected to the grid.

It improves the integration of renewable energy sources by ensuring effective power transmission and system stability as given in Table 8-9 and Fig. 9. The mean output of the inverter is 126 kW, capacity factor is 0.252%and losses are 58,038 kWh/year.

*Table 8. Converter power output for grid utilization*

| Quantity | Inverter |
|---|---|
| Power capacity (kW) | 50,000 |
| Mean output power (kW) | 126 |
| Minimum output power(kW) | 0 |
| Maximum output power(kW) | 646 |
| Capacity factor (%) | 0.252 |

*Table 9. Energy exchange by converter for better efficiency*

| Quantity | Inverter |
|---|---|
| Hours of operation (hrs/year) | 4,372 |
| Energy out (kWh/year) | 1,102,713 |
| Energy In (kWh/year) | 1,160,751 |
| Losses (kWh/year) | 58,038 |

*Figure 9. System converter output characteristics for agricultural power supply*

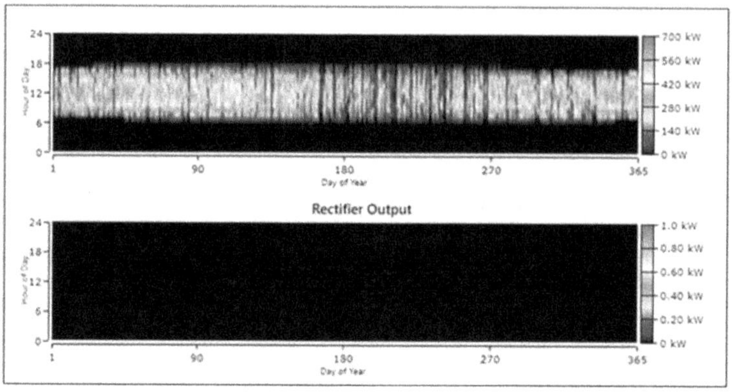


*(c). Grid Integrated Interfacing:* When demand exceeds supply, customers and the system trade energy by buying electricity and selling excess renewable energy. Credits for extra energy can be put back into the grid thanks to net metering. During times of low renewable output, reliance on the grid for power increases. In the energy markets, utilities make transactions easier. Technologies for smart grids maximize energy flow and encourage the incorporation of renewable. The dynamics in the changing energy environment emphasize sustainability and carbon reduction, with variations depending on local rules and resource availability as presented in Table 10. From the value below, it can analyse that the highest grid purchase is in made in the month of March which is 107,436 kWh while the maximum energy is also sold in the month of March which is 109,744 kWh. Vegetables like Cucumber, Cucumber, Bitter gourd, Gourd, Luffa, Petah, Spinach, Cauliflower, Brinjal, Okra, Arabic Mirchi Tomato are grown in the month of March which requires more energy for production so, the maximum energy is purchased in the month of March. The annual energy sold is 1,069,919 kWh.

*Table 10. Grid integrated power bidirectional supply*

| Month | Energy purchased (kWh) | Energy sold (kWh) | Net energy purchased (kWh) | Peak load (kW) | Energy charge ($) | Demand charge ($) |
|---|---|---|---|---|---|---|
| January | 2,692 | 101,625 | -98,933 | 20 | 0.00 | 0.00 |
| February | 2,080 | 101,331 | -99,251 | 17 | 0.00 | 0.00 |
| March | 2,309 | 109,744 | -107,436 | 19 | 0.00 | 0.00 |
| April | 2,204 | 103,463 | -101,259 | 19 | 0.00 | 0.00 |
| May | 2,159 | 96,303 | -94,143 | 19 | 0.00 | 0.00 |
| June | 2,013 | 71,017 | -69,004 | 19 | 0.00 | 0.00 |
| July | 2,024 | 61,511 | -59,486 | 19 | 0.00 | 0.00 |
| August | 2,319 | 61,598 | -59,279 | 18 | 0.00 | 0.00 |
| September | 2,233 | 69,599 | -67,366 | 18 | 0.00 | 0.00 |
| October | 2,448 | 93,113 | -90,664 | 17 | 0.00 | 0.00 |
| November | 2,492 | 97,715 | -95,223 | 19 | 0.00 | 0.00 |
| December | 2,618 | 102,900 | -100,282 | 18 | 0.00 | 0.00 |
| Annual | 27,592 | 1,069,919 | -1,042,327 | 20 | $103,190.42 | 0.00 |

*Figure 10. Characteristics of the energy purchased and sold through grid for agricultural activity*

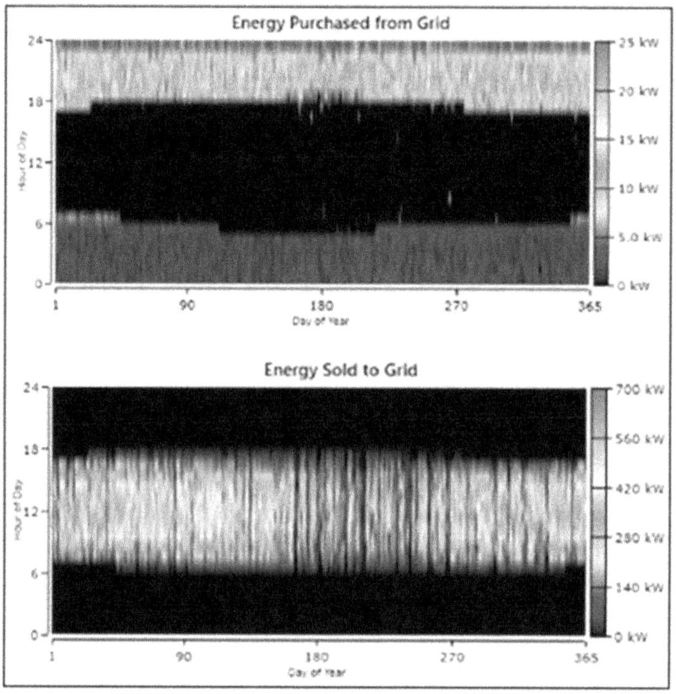


A grid purchase daily profile shows the amount of power purchased from the grid over the course of a day, highlighting demand variations. Fig. 10 shows the characteristics of the energy purchased and sold through grid while Fig. 11 shows the grid purchase daily profile. Demand peaks denote strong demand, whilst demand troughs denote low demand. Peaks often happen in the morning and evening, coinciding with everyday tasks like cooking and housekeeping. Recognizing these trends facilitates effective energy management.

*Figure 11. Grid purchase daily profile for agriculture energy supply*

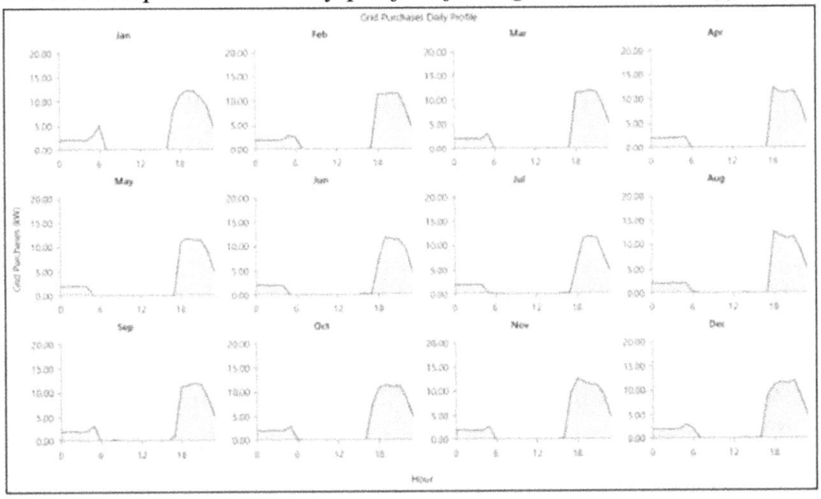


*Table 11. Results for renewable output by load for crop management*

| **Capacity based matrices for agrarian tasks** | **Value (%)** |
|---|---|
| Capacity based metrics | |
| Nominal renewable capacity divided by total nominal capacity | 100 |
| Usable renewable capacity divided by total capacity | 100 |
| Renewable output by load for crop management | |
| Peak values | |
| Renewable output divided by load (HOMER standard) | 105 |
| Renewable output divided by total generation | 100 |
| One minus non-renewable output divided by total load | 100 |
| Energy based analysis for rural activities | |
| Energy-based metrics | |
| Total renewable production divided by load | 103 |
| Total renewable production divided by generation | 97.7 |
| One minus total non-renewable production divided by load | 100 |

The amount or proportion of renewable energy in a certain energy system or grid is measured as renewable penetration. It shows how much of the total energy generated comes from renewable sources. When evaluating sustainability and environmental effect, this statistic is essential as shown in Table 11. Achieving greater energy independence, lowering dependency on fossil fuels, promoting climate change mitigation and moving toward cleaner energy all depend on increasing the share of renewable energy. Effective integration of renewable is necessary to achieve considerable penetration, overcoming issues like grid stability and intermittency.

The DC operational capacity daily profile shows how changes in system features, weather and sunshine affect a solar PV system's output throughout the day. The capacity increases with sunrise, reaches its maximum around solar noon and decreases with sunset. Fig. 12 shows the DC operating capacity daily profile characteristics. The average DC Operating capacity is calculated to be 300 kW. The profile is influenced by elements such as tracking systems and tilt kind of panel. Using solar simulation software and historical weather data, comprehensive modelling or monitoring is necessary to obtain an accurate profilee.

*Figure 12. Characteristics of the DC operating capacity daily profile*

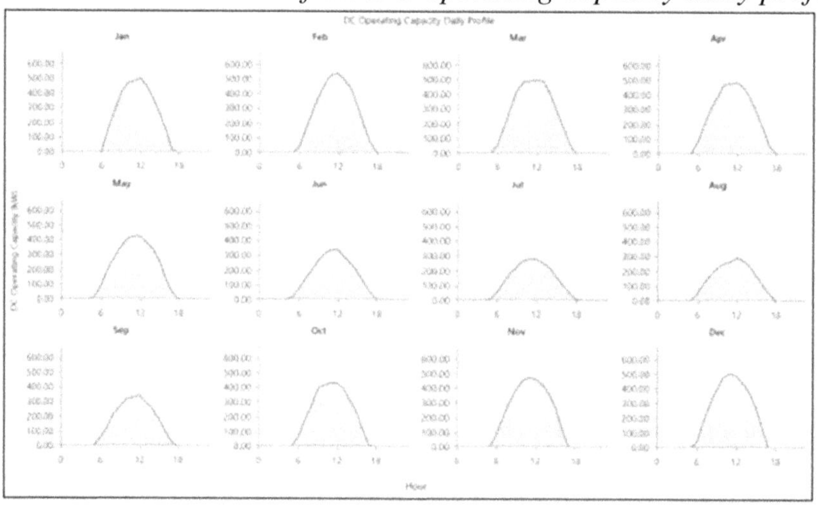

A solar PV system's daily profile for AC primary load and needed operational capacity is determined by a number of factors, such as system design, solar irradiation and patterns of energy consumption. Fig. 13 Showcases the AC required operating capacity daily profile characteristics. It's crucial to remember that the real profiles will rely a lot on the location's unique qualities, the kind of energy required and the efficiency and design of the solar PV system.

*Figure 13. Characteristics of AC required operating capacity daily profile*

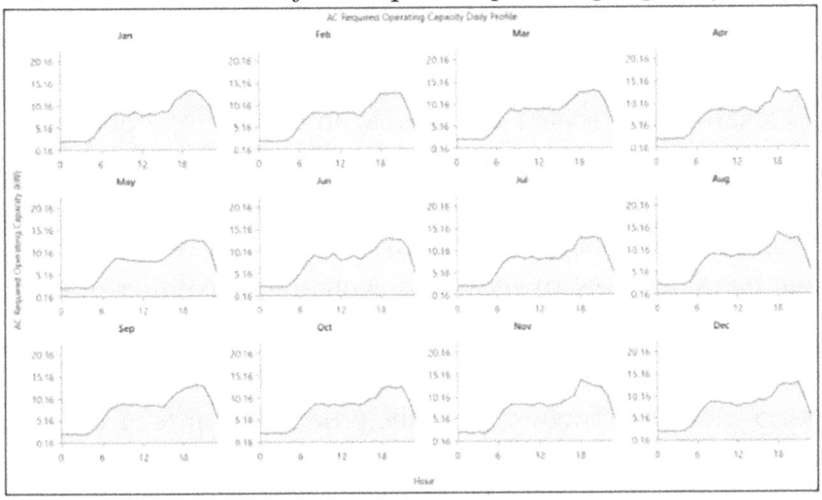


*(ii). Economic Viability Analysis:* A graphic depiction or table that shows varying rates or prices for various goods, services, or activities is called a rate chart as given in Table 12. It offers a methodical means of communicating expenses, levies, or fees related to certain goods or activities. Rate charts are frequently used to communicate with people in a variety of situations and sectors.

*Table 12. Rate chart comparison data analysis for farm operations*

| Financial Aspects | Data |
|---|---|
| Discount rate (%) | 8.00 |
| Inflation rate(%) | 2.00 |
| Real Interest Rate (%) | 5.9 |
| Project Lifetime(Years) | 25.00 |

Assessment of the initial investment assists in determining the expenses related to installing renewable energy sources, such solar or wind power. When weighing their investment alternatives, decision-makers must be aware of these expenses. Evaluation of operational costs makes it possible to look closely at the costs associated with continuing operations, such as upkeep, repairs and replacement parts for renewable energy systems. Examining the expenses associated with producing energy from renewable sources against conventional fossil fuel-based energy sources can provide light on the competitiveness and economic feasibility of renewable technologies.

Cost analysis assists by contrasting the initial investment with the anticipated savings or income from the production of renewable energy. This measure is crucial for determining if renewable energy projects are financially feasible and appealing. By knowing the financial effects of different government subsidies, policies and incentives pertaining to renewable energy, stakeholders may maximize financial gains and optimize their plans. The total cost of the system is 673,331.92$. To summarise, cost summary analysis is an essential component of renewable energy-based systems as it offers significant financial insights, facilitates well informed decision making and optimises the economic returns on renewable energy expenditures.

A company's cash inflows and outflows are shown by its cash flow characteristics, which are essential for comprehending liquidity and financial health. It reveals a variety of cash sources and uses by classifying transactions into operating, investing and financing activities. Analysis helps with financial planning by pointing up important drivers, cycles and trends. Comprehensive analyses augment lucidity however liquidity metrics provide discernment into immediate commitments. Fig. 14 and

Table 13 show, the cash flow characteristics under various conditions including by component and by nu discounted price. This forward-looking tool forecasts future financial positions based on anticipated trends and assists in strategic decision-making.

*Figure 14. Cash flow characteristic under various conditions for operation of the system*

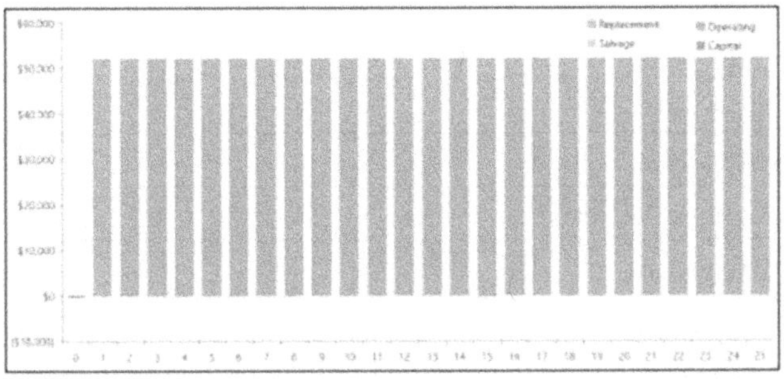

*Table 13. Economical comparison for the proposed system*

| Metric | Value |
|---|---|
| Simple payback (Year) | 4.58 |
| Return on Investment (%) | 17.8 |
| Internal Rate of Return (%) | 21.7 |
| Net Present Value ($) | 911,527 |
| Capital Investment ($) | 500,300 |
| Annualized Savings ($) | 109,211 |

*(iii). Environmental Impact Assessment:* Carbon dioxide emissions refer to the release of carbon dioxide gas into the atmosphere and they play a significant role in the context of climate change and environmental sustainability. The carbon dioxide produced here is 138 kg/year, the particular matter here is 166 kg/year, the sulphur dioxide produced here is 75.6 kg/year and nitrogen dioxide produced here is 37.0 kg/year as can be observed in Table 14. The emissions

here are basically produced because of the grid been installed as the electricity bought from the distribution network in grid-connected PV-storage microgrids may indirectly contribute to carbon dioxide emissions in the power grid.

*Table 14. Emissions reduction for environmental impact assessment*

| **Quantity** | **Value** |
|---|---|
| Carbon Dioxide (kg/year) | 414 |
| Carbon Monoxide (kg/year) | 0 |
| Unburned Hydrocarbons (kg/year) | 0 |
| Particulate Matter (kg/year) | 166 |
| Sulphur Dioxide (kg/year) | 75.6 |
| Nitrogen Oxides (kg/year) | 37.0 |

This outcome refer to Fig. 15 indicates the precise values given to the decision variables in order to maximize the objective function and guarantee the best possible performance or efficiency within the parameters of the problem under consideration. In Table 15, the NPC calculated using the formula is -$833,463and COE is -$0.0570. The annual operating cost is -$103,172. The initial capital for setting up the system is $500,300and the renewable fraction is 97.6%.

*Table 15. Optimization result for energy supply to agricultural requirements*

| | | | Schn 680 (kW) | Schn 680-MPPT (kW) | Grid (kW) | Converter (kW) | Dispatch | NPC ($) | COE ($) | Operating Cost ($/year) | Initial Capital ($) | Renewable Fraction (%) |
|---|---|---|---|---|---|---|---|---|---|---|---|---|
| | | | 680 | 680 | 999,999 | 50,000 | CC | -$833,463 | -$0.0570 | -$103,172 | $500,300 | 97.6 |
| | | | | | 999,999 | | CC | $78,064 | $0.100 | $6,039 | $0.00 | 0 |

*Figure 15. Comparative optimization characteristics for sustainable energy growth*

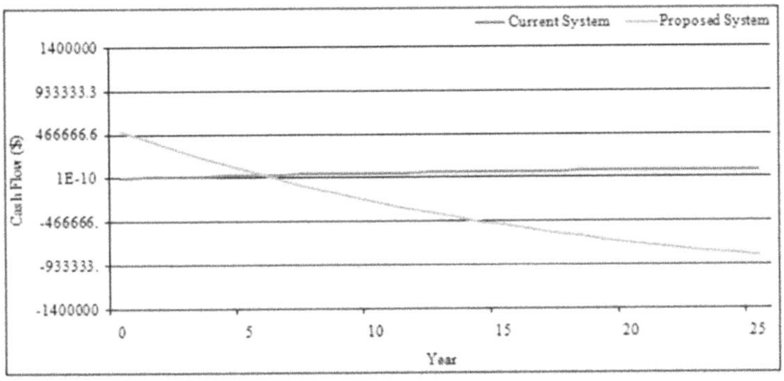


## CONCLUSION

Grid analysis assesses the efficiency and dependability of electrical networks, taking into account demand patterns, load distribution, and renewable energy integration. Similarly, PV analysis evaluates the efficacy of solar power systems, concentrating on crucial aspects such as solar irradiation, panel orientation, and shading in order to maximize performance. PV power generation is the pivotal process of converting sunlight into electricity through photovoltaic cells. The global solar monthly and daily profiles encapsulate the intricate variation in solar energy availability worldwide. Dependent on factors such as Earth's tilt, latitude and climate conditions, these profiles provide a holistic view of solar energy patterns. The total energy produced by the PV panels is 1,160,751 kWh/year. Simulation tools further refine the accuracy of these profiles, offering location-specific insights crucial for the development of sustainable energy solutions. In essence, the understanding of these diverse facets is imperative for navigating the complex landscape of renewable energy, emissions reduction and the evolving dynamics of global energy systems. In conclusion to this the present worth of the proposed system is estimated to be $751,396 while the annual worth is $58,124. The return on investment for the project is 19,378.35% just because of the energy being sold to the grid but if the proposed system is utilized for the sustainable agriculture practice it will find its worth. The carbon dioxide produced here is 138 kg/ year. Thus there are no health hazards caused due to the proposed system.

## REFERENCES


Abot, M. D. (2020). Factors influencing adoption of artificial insemination by smallholder livestock farmers in dryland production systems of Kenya *(Doctoral dissertation, University of Nairobi)*.

Bag, S., Yadav, G., Dhamija, P., & Kataria, K. K. (2021). Key resources for industry 4.0 adoption and its effect on sustainable production and circular economy: An empirical study. *Journal of Cleaner Production, 281*, 125233. DOI: 10.1016/j.jclepro.2020.125233

Bharathi, K., & Sasikumar, M. (2021). Power flow control based on bidirectional converter for hybrid power generation system using microcontroller. *Microprocessors and Microsystems, 82*, 103950. DOI: 10.1016/j.micpro.2021.103950

Bhuiyan, M. R. A. (2022). Overcome the future environmental challenges through sustainable and renewable energy resources. *Micro & Nano Letters, 17*(14), 402–416. DOI: 10.1049/mna2.12148

Braun, K., & Kropp, C. (2023). Building a better world? Competing promises, visionsand imaginaries-in-the-making of the digitalization of architecture and construction. *Futures, 154*, 103262. DOI: 10.1016/j.futures.2023.103262

Chauhan, C., Parida, V., & Dhir, A. (2022). Linking circular economy and digitalisation technologies: A systematic literature review of past achievements and future promises. *Technological Forecasting and Social Change, 177*, 121508. DOI: 10.1016/j.techfore.2022.121508

Chel, A., & Kaushik, G. (2011). Renewable energy for sustainable agriculture. *Agronomy for Sustainable Development, 31*(1), 91–118. DOI: 10.1051/agro/2010029

Chowdhury, S., Dey, P. K., Rodríguez-Espíndola, O., Parkes, G., Tuyet, N. T. A., Long, D. D., & Ha, T. P. (2022). Impact of organisational factors on the circular economy practices and sustainable performance of small and medium-sized enterprises in Vietnam. *Journal of Business Research, 147*, 362–378. DOI: 10.1016/j.jbusres.2022.03.077

de Oliveira, R. T., Ghobakhloo, M., & Figueira, S. (2023). Industry 4.0 Towards Social and Environmental Sustainability in Multinationals: Enabling Circular Economy, Organizational Social Practices and Corporate Purpose. *Journal of Cleaner Production*, *430*, 139712. DOI: 10.1016/j.jclepro.2023.139712

Dillon, E. J., Hennessy, T., & Hynes, S. (2010). Assessing the sustainability of Irish agriculture. *International Journal of Agricultural Sustainability*, *8*(3), 131–147. DOI: 10.3763/ijas.2009.0044

Furstenau, L. B., Sott, M. K., Kipper, L. M., Machado, E. L., Lopez-Robles, J. R., Dohan, M. S., Cobo, M. J., Zahid, A., Abbasi, Q. H., & Imran, M. A. (2020). Link between sustainability and industry 4.0: Trends, challenges and new perspectives. *IEEE Access : Practical Innovations, Open Solutions*, *8*, 140079–140096. DOI: 10.1109/ACCESS.2020.3012812

Gatell, I. S., & Avella, L. (2024). Impact of Industry 4.0 and circular economy on lean culture and leadership: Assessing digital green lean as a new concept. *European Research on Management and Business Economics*, *30*(1), 100232. DOI: 10.1016/j.iedeen.2023.100232

Guan, L., Li, W., Guo, C., & Huang, J. (2023). Environmental strategy for sustainable development: Role of digital transformation in China's natural resource exploitation. *Resources Policy*, *87*, 104304. DOI: 10.1016/j.resourpol.2023.104304

Han, H., & Zhou, Z. (2024). The rebound effect of energy consumption and its determinants in China's agricultural production. *Energy*, *290*, 129961. DOI: 10.1016/j.energy.2023.129961

Hassan, Q., Sameen, A. Z., Salman, H. M., & Jaszczur, M. (2023). Large-scale green hydrogen production via alkaline water electrolysis using solar and wind energy. *International Journal of Hydrogen Energy*, *48*(88), 34299–34315. DOI: 10.1016/j.ijhydene.2023.05.126

Hettiarachchi, B. D., Brandenburg, M., & Seuring, S. (2022). Connecting additive manufacturing to circular economy implementation strategies: Links, contingencies and causal loops. *International Journal of Production Economics*, *246*, 108414. DOI: 10.1016/j.ijpe.2022.108414

James, A. T., Kumar, G., James, J., & Asjad, M. (2023). Development of a micro-level circular economy performance measurement framework for automobile maintenance garages. *Journal of Cleaner Production, 417*, 138025. DOI: 10.1016/j.jclepro.2023.138025

Jat, H. S., Datta, A., Choudhary, M., Sharma, P. C., & Jat, M. L. (2021). Conservation Agriculture: Factors and drivers of adoption and scalable innovative practices in Indo-Gangetic plains of India–a review. *International Journal of Agricultural Sustainability*, *19*(1), 40–55. DOI: 10.1080/14735903.2020.1817655

Knickel, K., Ashkenazy, A., Chebach, T. C., & Parrot, N. (2017). Agricultural modernization and sustainable agriculture: Contradictions and complementarities. *International Journal of Agricultural Sustainability*, *15*(5), 575–592. DOI: 10.1080/14735903.2017.1373464

Koohafkan, P., Altieri, M. A., & Gimenez, E. H. (2012). Green agriculture: Foundations for biodiverse, resilient and productive agricultural systems. *International Journal of Agricultural Sustainability*, *10*(1), 61–75. DOI: 10.1080/14735903.2011.610206

Kurniawan, T. A., Maiurova, A., Kustikova, M., Bykovskaia, E., Othman, M. H. D., & Goh, H. H. (2022). Accelerating sustainability transition in St. Petersburg (Russia) through digitalization-based circular economy in waste recycling industry: A strategy to promote carbon neutrality in era of Industry 4.0. *Journal of Cleaner Production*, *363*, 132452. DOI: 10.1016/j.jclepro.2022.132452

Kurniawan, T. A., Othman, M. H. D., Hwang, G. H., & Gikas, P. (2022). Unlocking digital technologies for waste recycling in Industry 4.0 era: A transformation towards a digitalization-based circular economy in Indonesia. *Journal of Cleaner Production*, *357*, 131911. DOI: 10.1016/j.jclepro.2022.131911

Kurniawan, T. A., Othman, M. H. D., Liang, X., Goh, H. H., Gikas, P., Kusworo, T. D., Anouzla, A., & Chew, K. W. (2023). Decarbonization in waste recycling industry using digitalization to promote net-zero emissions and its implications on sustainability. *Journal of Environmental Management, 338*, 117765. DOI: 10.1016/j.jenvman.2023.117765 PMID: 36965421

Liu, Y., Yu, Y., Huang, Y., & Guan, W. (2024). Utilizing the resources efficiency: Evidence from the impacts of media industry and digitalization. *Resources Policy, 88*, 104346. DOI: 10.1016/j.resourpol.2023.104346

Matt, D. T., Pedrini, G., Bonfanti, A., & Orzes, G. (2023). Industrial digitalization. A systematic literature review and research agenda. *European Management Journal, 41*(1), 47–78. DOI: 10.1016/j.emj.2022.01.001

Mint, (August 14, 2023). *Indian agriculture must balance profits with sustainable practices*.https://www.livemint.com/opinion/first-person/indian-agriculture-must-balance-profits-with-sustainable-practices-11691889108165.html (Accessed on: December 18, 2023).

Modgil, S., Gupta, S., Sivarajah, U., & Bhushan, B. (2021). Big data-enabled large-scale group decision making for circular economy: An emerging market context. *Technological Forecasting and Social Change, 166*, 120607. DOI: 10.1016/j.techfore.2021.120607

Munir, M. T., Li, B., Naqvi, M., & Nizami, A. S. (2023). Green loops and clean skies: Optimizing municipal solid waste management using data science for a circular economy. *Environmental Research*, •••, 117786. PMID: 38036215

Pal, P., & Mukherjee, V. (2021). Off-grid solar photovoltaic/hydrogen fuel cell system for renewable energy generation: An investigation based on techno-economic feasibility assessment for the application of end-user load demand in North-East India. *Renewable & Sustainable Energy Reviews, 149*, 111421. DOI: 10.1016/j.rser.2021.111421

Pretty, J., Smith, G., Goulding, K. W. T., Groves, S. J., Henderson, I., Hine, R. E., King, V., Van Oostrum, J., Pendlington, D. J., Vis, J. K., & Walter, C. (2008). Multi-year assessment of Unilever's progress towards agricultural sustainability II: Outcomes for peas (UK), spinach (Germany, Italy), tomatoes (Australia, Brazil, Greece, USA), tea (Kenya, Tanzania, India) and oil palm (Ghana). *International Journal of Agricultural Sustainability*, *6*(1), 63–88. DOI: 10.3763/ijas.2007.0323

Ritala, P., Baiyere, A., Hughes, M., & Kraus, S. (2021). Digital strategy implementation: The role of individual entrepreneurial orientation and relational capital. *Technological Forecasting and Social Change*, *171*, 120961. DOI: 10.1016/j.techfore.2021.120961

Sharma, H. B., Vanapalli, K. R., Samal, B., Cheela, V. S., Dubey, B. K., & Bhattacharya, J. (2021). Circular economy approach in solid waste management system to achieve UN-SDGs: Solutions for post-COVID recovery. *The Science of the Total Environment*, *800*, 149605. DOI: 10.1016/j.scitotenv.2021.149605 PMID: 34426367

Sun, L., Li, X., & Wang, Y. (2024). Digital trade growth and mineral resources In developing countries: Implications for green recovery. *Resources Policy*. 88, pp. 104338. S&P Global, (August 3, 2023). *Future Farming: Agriculture's Role in a More Sustainable India*. https://www.spglobal.com/en/research-insights/featured/special-editorial/look-forward/future-farming-agriculture-s-role-in-a-more-sustainable-india (Accessed on: December 18, 2023).

The News India Express. (August 18, 2023). *India's agriculture sustainability at risk*. https://www.newindianexpress.com/xplore/2023/aug/18/indiasagriculture-sustainability-at-risk-2606814.html (Accessed on: December 18, 2023).

The Sunday Guardian. (October 14, 2023). *The Green Revolution in India: Harnessing Electric Tractors for Sustainable Agriculture*. https://sundayguardianlive.com/business/the-green-revolution-in-india-harnessing-electric-tractors-for-sustainable-agriculture (Accessed on: December 18, 2023).

The Times of India. (May 4, 2023). *Embracing sustainable agriculture: A virtuous cycle of prosperity*. https://timesofindia.indiatimes.com/blogs/voices/embracing-sustainable-agriculture-a-virtuous-cycle-of-prosperity/ (Accessed on: December 18, 2023).

The Wire. (July 21, 2023). *How Natural Farming Can Revive India's Farmlands and Ensure Sustainable Agriculture*.https://thewire.in/agriculture/how-natural-farming-can-revive-indias-farmlands-and-ensure-sustainable-agriculture (Accessed on: December 18, 2023).

Truant, E., Giordino, D., Borlatto, E., & Bhatia, M. (2024). Drivers and barriers of smart technologies for circular economy: Leveraging smart circular economy implementation to nurture companies' performance. *Technological Forecasting and Social Change*, *198*, 122954. DOI: 10.1016/j.techfore.2023.122954

Upadhyay, A., Balodi, K. C., Naz, F., Di Nardo, M., & Jraisat, L. (2023). Implementing industry 4.0 in the manufacturing sector: Circular economy as a societal solution. *Computers & Industrial Engineering*, *177*, 109072. DOI: 10.1016/j.cie.2023.109072

Vatican News. (December 04, 2023). *COP28: Sustainable agriculture is antidote to climate change*.https://www.vaticannews.va/en/world/news/2023-12/cop28-sustainable-agriculture-antidote-to-climate-change.html (Accessed on: December 18, 2023).

Wang, Z., Huang, Y., Ankrah, V., & Dai, J. (2023). Greening the knowledge-based economies: Harnessing natural resources and innovation in information and communication technologies for green growth. *Resources Policy*, *86*, 104181. DOI: 10.1016/j.resourpol.2023.104181

Zhang, H., & Xie, Y. (2024). Assessing natural resources, rebounding trends, digital economic structure and green recovery dynamics in China. *Resources Policy*, *88*, 104482. DOI: 10.1016/j.resourpol.2023.104482

## KEY TERMS AND DEFINITIONS

**ABARE:** Australian Bureau of Agricultural and Resource Economics.

**AM:** Additive Manufacturing.

**AN:** Agricultural Nutrients.

**API:** Application Programming Interfaces.

**ASEAN:** Association of Southeast Asian Nations.

**BDA:** Big Data Analytics.

**BPNA:** British Paediatric Neurology Association.

**BRICS:** Brazil, Russia, India, China and South Africa.

**CA:** Conservation Agriculture.

**CE:** Circular Economy.

**CIAS:** Composite Index of Agricultural Sustainability.

**COP28:** 28th United Nations Climate Change Conference.

**CWA:** Cost and Works Accountant.

**DT:** Digital Transformation.

**EEP:** Educational Effectiveness Plan.

**EIA:** Environmental Impact Assessments.

**ES:** Environmental Strategy.

**FAO:** Food and Agriculture Organization.

**GAMC:** Global Conference on Sustainable Agricultural Mechanization.

**GHI:** Global Horizontal Irradiance.

**IEO:** Innovative Entrepreneurial Orientation.

**IGP:** Indo-Gangetic Plain.

**ICTs:** Information and Communication Technologies.

**IOT:** Internet of Things.

**ISC:** Closed Circuit Current.

**KBEs:** Knowledge-Based Economies.

**MNEs:** Multinational Enterprises.

**MSW:** Municipal Solid Waste.

**NAM:** National Agriculture Market.

**NPC:** Net Present Cost.

**NR:** Natural Resource.

**PLS-SEM:** Partial Least Squares Structural Equation Modelling.

**PFDCs:** Precision Farming Development Centres.

**REMs:** Rare Earth Metals.

**SAPs:** Sustainable Agricultural Practices.

**SD:** Sustainable Development.

**SDGs:** Sustainable Development Goals.

**SLR:** Systematic Literature Review.

**SMEs:** Small and Medium Enterprises.

**SSAFNS:** Smart & Sustainable Agriculture for Food and Nutritional Security.

**UN:** United Nations.

**VOC:** Open Circuit Voltage.

**WoS:** Web of Science.

# Chapter 7
# Optimization for Unified Production of Electricity and Green Hydrogen With Hybrid Renewable Energy Sources

**Raju Kumar Maurya**
*Department of Electrical and Electronics Engineering, SRM Institute of Science and Technology, Modinagar, India*

**Dharmbir Prasad**
https://orcid.org/0000-0002-9010-9717
*Department of Electrical and Electronics Engineering, SRM Institute of Science and Technology, Modinagar, India*

**Rudra Pratap Singh**
https://orcid.org/0000-0001-7352-855X
*Department of Electrical Engineering, Asansol Engineering College, Asansol, India*

**Ranjeet Singh**
*Department of Electrical Engineering, Faculty of Technology, Delhi University, India*

**Aditya Arya**
*Department of Electrical Engineering, Asansol Engineering College, Asansol, India*

**Ranadip Roy**
https://orcid.org/0000-0003-2111-2581
*Department of Electrical Engineering, Sanaka Educational Trust's Group of Institutions, Durgapur, India*

DOI: 10.4018/979-8-3693-0827-1.ch007

**Sudhangshu Chakraborty**
https://orcid.org/0000-0002-4874-9681

*Department of Basic Science and Humanity (Physics), Asansol Asansol Engineering College, Asansol, India*

## ABSTRACT

*The present work focused on green hydrogen production in microgrids as a sustainable and reliable energy source for off-grid areas. Green hydrogen is produced using renewable energy sources, offers a clean and efficient alternative to fossil fuels. This study explores technical feasibility and economic viability for unified production of green hydrogen and electricity. In the proposed microgrid model, the energy production and consumption pattern of residing community is analyzed. The number of factors such as size and capacity of the hydrogen production and storage systems are evaluated to find reliable and economic integrated system. It is found to be 450 kg/year green hydrogen output from the proposed Chintamanpur site of Bihar state. It provide reliable access to clean energy while minimizing greenhouse gas emissions and increasing energy security. The study also highlights importance of policy support and investment in RE infrastructure to enable the widespread adoption of microgrid based hydrogen production unit as a key component of the sustainable energy transition.*

## INTRODUCTION

A potential substitute for facilitating an ecofriendly energy shift is green hydrogen. In addition to replacing fossil fuels in industry and providing clean transportation, hydrogen can also be used for decentralized power generation, aviation, and maritime transportation. India is one of the fastest developing economies with a population of over 140.76 crore people (UN DESA, 2023). Despite its efforts to increase access to electricity, a major section of the population still lacks reliable and affordable energy. In the last year, India increased its nationally decided contribution, moving closer to its objective of having net-zero emissions by the year 2070 (Ministry of Environment, Forest and Climate Change, 2023). But most of the hydrogen produced today is the 'black' or 'brown' hydrogen, produced from coal. Grey hydrogen is produced from natural gas, whereas 'blue' hydrogen is made from fossil fuels with techniques that ensure the carbon exhaled is collected (Huang et al., 2023). By 2050, green hydrogen could reduce $CO_2$ emissions by 3.6 gigatons (NITI Aayog, 2022). It is formed by water electrolyzationlike, when electricity is produced using RE sources like solar, wind, etc.,and further, water is split into hydrogen and oxygen. This method of hydrogen production is the most environmentally friendly with the hybrid RE solution portrayed in Fig. 1.

Several industries, including heavy industry (*e.g.,* ammonia, steel and refinery), aviation and power generation, will drive this growth. There will be a 3.3 fold rise in the global demand for hydrogen by 2050. According to the analysis (S&P Global Commodities Insights, 2023), 67% of the $H_2$ will be green, with 16% (40MT) of it being blue (produced from fossil fuels with carbon capture and storage). The demand for hydrogen in India is anticipated to grow by 2.5-3.5 times by 2040; however, it is still not anticipated to account for more than 5% of the country's total primary energy consumption (Asian Development Bank, 2023). To meet its growing electricity demands and reduce its reliance on fossil fuels, India has made huge financial investments in RE sources such as solar and wind power (The Indian Express, 2023). However, the intermittent nature of these sources presents a challenge for ensuring reliable access to energy. Green hydrogen production in microgrids has emerged as a promising solution to address this challenge. These microgrids use RE sources to produce green hydrogen through electrolysis, which can be stored and used to generate electricity whenever needed (Qiu et al., 2023). It has the potential to provide a reliable and sustainable source of energy for remote and off-grid communities, while also reducing carbon emissions.

*Figure 1. Schematic for selection interconnection of optimal hybrid RE system*

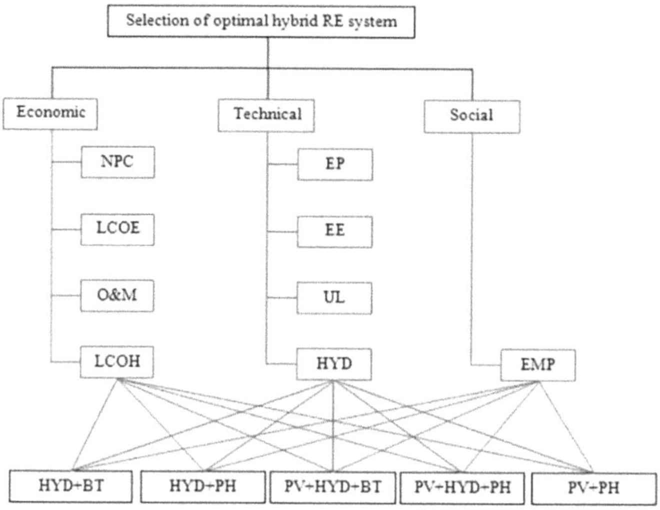


This study presents the potential benefits and challenges of implementing green hydrogen-basedmicrogrids. It highlights the application of technology to increase energy access, reduce carbon emissions, and improve energy security. Globally, the development of green hydrogen is still in its nascent stage, and while India can lead the way by becoming a significant producer, it now lacks the infrastructure required to carry out all these intermediary phases. Incentives must also be provided to persuade enough industrial hydrogen users to switch to green hydrogen. To ensure that lakhs of people can be properly trained to adapt to a viable green hydrogen economy, it must construct supply chains in the form of pipelines, tankers, intermediate storage, and last-leg distribution networks. It must also put in place an effective skill development program. Additionally, it highlights the challenges, including high upfront costs, limited infrastructure, and regulatory barriers(The Energy and Resources Institute, 2023). Despite these challenges, with the right policies and investments, green hydrogen production has the potential to transform India's energy landscape, providing reliable and sustainable energy to its growing population with a lower carbon footprint.

The rest of the paper is organized inthe following sequence: problem statement, background review, mathematical modeling, numerical analysis approach, site potential and system capacity sizing, system modeling, the results, and discussion. Finally, the paper has been concluded.

## PROBLEM STATEMENT

Green hydrogen production using excess energy in microgrids is recognized as a possible solution to uncertainty and excess energy issues. However, the high upfront costs and limited infrastructure for this technology present significant challenges for its widespread adoption. In the present work, the following objectives for green hydrogen production in a microgrid are attained:

- Renewable penetration and excess electricity utilization
- Techno-economic analysis for the unified model
- Emission reduction with green hydrogen production

## BACKGROUND REVIEW

Energy transition is focused on maximizing the usage of renewable energy across all economic sectors. Recent research to facilitate this shift are green hydrogen are summaries as below:

*(i). Renewable system integration:* Towards green energy initiative, a study demonstrates a flexible hybrid energy installation for the Shiraz Elam Campus that integrates renewable energy such as electricity and heat (Eisapour et al., 2021). Further, the study suggested usage of waste energy, especially waste electricity, can reduce energy costs by 18% (HassanzadehFard et al., 2023). Hospitals can use HRES and hydrogen products to power the COVID-19 ward, which reduces CO2 emissions and provides energy and oxygen (Izadi et al., 2023).

*(ii). Renewable to hydrogen production:* The advancement in the generation of renewable storage-based fuels such as hydrogen and lithium-ion batteries is transforming the renewable power generation model (Zhang et al., 2020). Research in renewable energy optimization techniques using clean storage systems along with microgrid systems offers efficient and environmental solutions (Akarsu & Genç, 2022). The importance of renewable energy systems together with hydrogen in solving energy-related problems related to optimization, and production process (Khan et al., 2022). A two-stage development system for decision makers, further, supports hydrogen refueling and carbon trading (He et al., 2022). The availability of hydrogen has improved the integration of HRES into zero buildings to achieve energy independence (Izadi et al., 2022). The optimization of hydrogen energy usage can be achieved by integration of hydrogen into renewable energy sources with systematic management and time of day uses of the energy (Liu et al., 2022). Modeling captures changes and guarantees cost reduction and CO2 emissions. The result is that hydrogen can reduce emissions, albeit at a higher cost (Kilic et al., 2022).

Electric vehicles have advantages over hydrogen cars, combining hybrid energy storage for carbon-neutral buildings and transportation with new energy vehicles, improving design and reducing emissions and costs (Fan et al., 2022). Partnerships can improve HSS outcomes by emphasizing the importance of performance metrics for improvement (Luo et al., 2023). In another work, aims to create a zero-emissions energy system in the park by combining solar PV, wind, power-to-gas, and carbon capture to reduce emissions, create hydrogen/natural gas and diversify carbon sources (Jia et al., 2023). Next, optimization methods are used to reduce the energy and total cost of renewable energy to provide reliable electricity and heat for business needs (Azad & Shateri, 2023).

The optimization strategies for RE microgrids improve hybrid performance (Huangfu et al., 2023). Research on fuel oxidizers and biomass gasification is relevant for large-scale power generation and the low hydrogen content of syngas (Hai et al., 2023). There is an effort in Texas research to replace coal and natural gas with cost-effective N-RHES that combines nuclear, wind, solar, and green hydrogen to meet energy needs and promote clean business (Bryan et al., 2023). Then, a method is proposed for solving problems in renewable energy systems by integrating various devices (Yan et al., 2023). Then, the efficiency of green hydrogen production and the installed capacity of wind and solar power are integrated to construct a comprehensive indicator (Zhang et al., 2023). P2G can convert excessive renewable energy into hydrogen via electrolysis, which can then be transported by natural gas systems to bypass constrained electricity systems (Zhao et al., 2020). This article proposes a multi renewable-to-hydrogen production method to enhance the green $H_2$ production efficiency for renewable-dominated HFSs (Zhang et al., 2021). Hydrogen synthesis with the greenhouse gases to form hydrocarbons and reuse the fuel generated with electrolysis. In the IEPU framework, a PV power plant is installed to produce electricity; CO2 capture technology is applied to the existing coal-fired power plant with biomass co-combustion (Zhou et al., 2021).

A multi-energy trading framework for a HP has been proposed to coordinate the interaction and trading of electricity and H2 while promoting the efficient accommodation of RE resources (Zhang et al., 2022). Further, based on the clearing solution, the concept and calculation method of the LMHP are proposed concerning the LMP in electricity markets (Zheng et al., 2022). Green hydrogen in FCVs is a promising alternative in the transportation sector. However, high hydrogen costs hinder progress (Zheng et al., 2022). An OHRS can avoid hydrogen transportation and participate in the market to reduce costs. In the forward market, it can lock in inexpensive electricity in advance; meanwhile, lower prices can be attained in the spot market (Wu et al., 2022). An electricity price control mechanism is used for the operation of an electric thermal energy storage system (Bachmann et al., 2023).

*(iii). Green energy economic reviews:* Research on the optimization of off-grid hybrid renewable energy systems to meet electricity, heat, and hydrogen needs have been studied (Akhtari & Baneshi, 2019). The energy model for buildings in Spain from 2030 to 2050 combines solar, wind, and hydrogen technologies using geospatial optimization for low-cost decarbonization of energy consumption (Maestre et al., 2022). Another research was designed in school to create an integrated zero-emissions energy system that reduces

emissions while producing hydrogen and natural gas from a variety of carbons, (He et al., 2022). Next, a study optimizes photovoltaic-hydrogen hybrid systems to be carbon neutral to maximize profits from photovoltaic physics models and government subsidies and is useful to policymakers, business owners, and investors (Yang et al., 2023). In its continuation, a solar-geothermal cogeneration system for power, hydrogen, and freshwater production, achieving high efficiency to generate freshwater and economic benefits (Shahverdian et al., 2023). Integration of RE reduces the cost of electricity, hydrogen, and heat required for grid connection and distribution of electricity (Di Micco et al., 2023).

## MATHEMATICAL MODELLING

Green hydrogen production is done through water electrolysis, an endothermic chemical process that splits water ($H_2O$) into hydrogen ($H_2$) and oxygen ($O_2$) through the electric energy as $2H_2O \rightarrow 2H_2 + O_2$. This requires water, electricity, and heat as inputs, while electricity required to produce green hydrogen can be calculated with the following levelized cost formulation as given in (1) (Akhtari & Baneshi, 2019), (Bryan et al., 2023),

$$LCOE = \frac{\sum_{t=1}^{n} \frac{CAPEX_t + OPEX_t + F_t}{(1+r)^t}}{\sum_{t=1}^{n} \frac{E_t}{(1+r)^t}} \tag{1}$$

where, *CAPEX*: capital expenditure, *OPEX*: operational expenditure, *F*: fuel expenditure, *E*: electricity generation, *r*: discount rate, *n*: lifetime of the system in years, and *t*: value with the subscript *t* are relevant to the year.

In this study, it assumed that no fossil fuel will be consumed in case of availability of renewable power, so $F_t$ is set to zero. The discount rate is assumed to be constant over the time horizon. Whereas, the $CAPEX_t$ and $OPEX_t$ are calculated as the annual expenditure based on installation power capacity, and the generated electricity is assumed to be negligible depreciation over the lifetime of the system. Furthermore, considering all these techno-commercial aspects expression (1) can be simplified as in (2) (Akhtari & Baneshi, 2019), (Bryan et al., 2023),

$$LCOE = \frac{\frac{CAPEX}{PVF} + OPEX}{E} \tag{2}$$

However, the present value factor for this study can be formulated as in (3) (He et al., 2022),

$$PVF = \frac{(1+i)^t - 1}{(1+i)^t . i} \tag{3}$$

where, $i$: interest rate, which is taken to be 8% for all components of RE integration in this study.

The RE generation has been focused in this work to provide greener means of hydrogen production; mainly with PV, wind, and battery storage integration into the system.In this study, low-cost flat PV panels taken for RE system integration and corresponding power output can be given by (4) (Akhtari & Baneshi, 2019), (Bryan et al., 2023),

$$P_{pv}(t) = \left[PV_r \times PV_{df}\left(\frac{1}{I_{stc}}\right)\left\{1 + \alpha_p\left(T(t) - T_{stc}\right)\right\}\right] \tag{4}$$

where, $P_{pv}(t)$: solar output power at the time $t$, $PV_r$: power rating of the PV panel, $PC_{df}$: PV loss factor, $I_{stc}$: solar radiation incident, $\alpha_p$: temperature-coefficient for solar power generation, $T(t)$: actual temperature on the module and $T_{stc}$: standard temperature of the PV module.

In continuation of equation (1), further, the operating temperature can be expressed as in (5) (He et al., 2022),

$$T(t) = T_a(t) + \left\{\left(T_{c\_notc} - T_{a\_notc}\right)\left(\frac{1}{I_{stc}}\right)\left(1 - \frac{\eta_c}{\alpha . \tau}\right)\right\} \tag{5}$$

where, $T_a(t)$: ambient temperature (°C), $T_{c\_notc}$: specified operating temperature, $T_{a\_notc}$: ambient temperature condition, $\eta_c$: efficiency of the solar cell, $\alpha$: absorption of solar radiation and $\tau$: solar energy transmittance.

In case of weather intermittency and due to the lack of or limited availability of solar radiation at night and on cloudy days, WT may be utilized in conjunction with PV for dependable power generation while using (6) (Kushwaha & Bhattacharjee, 2022), (Pires et al., 2023),

$$P_w(t) = \begin{cases} 0 & v(t) < v_c \text{ or } v(t) > v_f \\ P_{rw}\left\{\dfrac{v(t) - v_c}{v_r - v_c}\right\} & v_c < v(t) < v_r \\ P_{rw} & v_r < v(t) < v_f \end{cases} \quad (6)$$

where, $P_w(t)$: wind power output at the time $t$, $P_{rw}$: rated wind power, $v(t)$: wind speed at a time $t$ for the undertaken site, $v_r$: rated speed of wind turbine, $v_c$: wind turbines cut out speed and $v_f$: wind turbines cut-in speed.

Due to the dynamic nature of weather, both PV and WT generation might be uncertain; then, the battery storage must therefore deliver energy in high-load conditions and absorb energy in low-load situations. Equation (7) is used to calculate the amount of energy that is currently available and (8) is used to estimate the battery's level of charge as given below (Kushwaha & Bhattacharjee, 2022), (Pires et al., 2023),

$$E_b(t) = E_b(t-1) + E_{ex}(t) \cdot \eta_{con} \cdot \eta_{bc} \quad (7)$$

$$SOC_b(\%) = \frac{q_b}{q_{b\_m}} \times 100 \quad (8)$$

where, $E_b(t)$: battery energy storage availability at the time $t$, $E_{ex}(t)$: surplus energy, $\eta_{con}$: efficiency for power conversion, $\eta_{bc}$: charging efficiency of the battery, $SOC_b$: state of charge for battery, $q_b$: battery charge and $q_{b\_m}$: maximum charge of battery.

Considering different ways of RE generation, the power transition between AC and DC components is carried out with power converters. To account for allowable overloads, a converter's rating is more than the maximum power flow in both directions, which can be calculated by (9) - (10) (Kushwaha & Bhattacharjee, 2022), (Pires et al., 2023),

$$\eta^{con}_{dc/ac}\cdot\left[P_{pv}(t)+P_b(t)\right] \leq N_{con}.PSC_r \tag{9}$$

$$\eta^{con}_{ac/dc}\cdot\left[P_w(t)\right] \leq N_{con}.PSC_r \tag{10}$$

where, $\eta^{con}_{dc/ac}$, $\eta^{con}_{ac/dc}$: power conversion efficiency, $N_{con}$: number of converters and $PSC_r$: rated power of system converters.

Now, *LCOH* attributing to the total specific hydrogen production cost can be formulated by (11) (Akhtari & Baneshi, 2019), (Bryan et al., 2023),

$$LCOH = \frac{\frac{CAPEX + c_{stack}}{PVF} + OPEX}{FLH.\frac{\eta_{ely}}{\omega_{H_2}}} + LCOE.\left(\frac{\omega_{H_2}}{\eta_{ely}}\right) + c_{H_2O}.v_{H_2O} \tag{11}$$

where, $c_{stack}$: cost of stack replacement, *FLH*: water transport costs, $\eta_{ely}$: electrolysis system efficiency, $w_{H_2}$: gravimetric energy density, $v_{H_2O}$: water volume and $c_{H_2O}$: cost for water availability.

Water availability is a key ingredient for hydrogen production by water electrolysis. This can be calculated as in (12) (Akhtari & Baneshi, 2019), (Bryan et al., 2023),

$$c_{H_2o} = c_{H_2O_{spec}} + c_{H_2O_{transport}} + E_{H_2O}.LCOE \tag{12}$$

where, $c_{H_2O_{spec}}$: specific water cost, $c_{H_2O_{transport}}$: water transport costs and$E_{H_2O}$: water purification energy demand.

## NUMERICAL ANALYSIS APPROACH

Numerical analysis is an important aspect of green hydrogen production in a microgrid as it helps to determine the optimal sizing and operation of the different components of the system. Numerical analysis involves mathematical models and simulations to predict the system's behavior under different operating conditions and to identify areas for optimization. The following are some of the numerical analysis techniques applicable for hydrogen preparation in a microgrid:

i) *Load flow analysis:* It is used to determine the distribution of power flows in a microgrid. It involves modeling the different components of the system and simulating the power flows under different load conditions. It helps to determine the optimal sizing of the different components of the system and to identify areas for improvement.
ii) *Economic analysis:* It is used to determine the cost of the system and to identify the most cost-effective solution. It involves modeling the system and simulating the costs of different components and operating scenarios. It also helps to identify the optimal sizing and operation of the different components of the system.
iii) *Environmental analysis:* It is used to determine the environmental impact of the system and to identify areas for improvement. It involves modeling the system and simulating the emissions under different operating conditions. It helps to identify the most sustainable solution.
iv) *Optimization techniques:* It is used to identify the optimal sizing and operation of the different components of the system. It involves mathematical algorithms to minimize the cost or maximize the performance of the system under compliance with different constraints.

## SITE POTENTIAL AND SYSTEM CAPACITY SIZING

The ideal location for a microgrid is in a remote or isolated area that is not connected to the main electricity grid. Finding a favorable microgrid location for green hydrogen production is an important factor in its design and optimization for long-term return. In this work, an industrial town (viz., Chintamanpur, coordinates - latitude: 25°57.9'N and longitude: 85°16.3'E) of Bihar state with abundant RE sources has been considered for coproduction of electricity and hydrogen. Climatic data plays an important role in the design and optimization of green hydrogen production. It includes key information such as temperature, solar radiation, wind speed, and humidity of the experimental zone as presented in Table 1 and Fig. 2. The monthly variation of average clearness index and solar radiation has been observed for the undertaken sites. In this work, the global horizontal solar radiation varies from 4.070 - 6.950 kWh/m$^2$/day, with a scaled annual average value of 5.08 kWh/m$^2$/day. In this study, ambient temperatures highly influence the performance of PV panels. However, the average wind speed over the year is 5.22 m/s. The GHI observation for solar radiation received by a surface horizontal to the ground is important for assessing the integrated solution for hydrogen production in microgrid systems.

*Table 1. RE potential of the undertaken site for the coproduction of electricity and green hydrogen*

| Month | Meteorological features | | | |
|---|---|---|---|---|
| | CI | DR | *T* | $N_w$ |
| January | 0.636 | 4.180 | 15.980 | 3.910 |
| February | 0.681 | 5.250 | 19.900 | 4.490 |
| March | 0.684 | 6.260 | 26.120 | 5.040 |
| April | 0.670 | 6.950 | 31.980 | 4.950 |
| May | 0.624 | 6.900 | 34.960 | 4.760 |
| June | 0.509 | 5.740 | 34.350 | 5.080 |
| July | 0.410 | 4.570 | 30.300 | 5.110 |
| August | 0.435 | 4.610 | 29.070 | 4.910 |
| September | 0.464 | 4.430 | 27.750 | 4.600 |
| October | 0.611 | 4.960 | 25.310 | 3.010 |
| November | 0.693 | 4.720 | 21.040 | 2.960 |
| December | 0.657 | 4.070 | 16.920 | 3.180 |

*Figure 2. Characteristics of RE generation meteorological parameters*

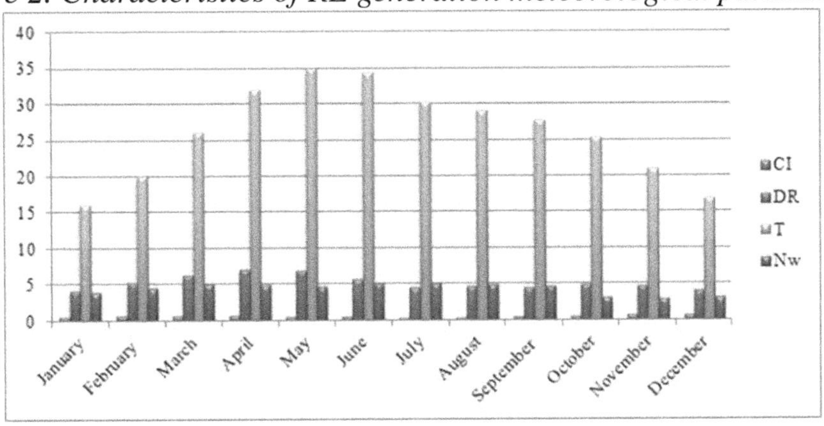

The present study of green hydrogen production in a microgrid aims to provide a reliable and sustainable source of energy using RE sources. This work was initiated with feasibility observations of the most suitable location and to assess the local energy needs and resources. Based on this analysis, the green hydrogen unit in a microgrid will be designed and integrated, including the necessary equipment for RE generation, hydrogen production, storage, and supply. This work will also serve in capacity building for the wide adoption of this technology requires training programs on the operation and maintenance of the equipment, as well as education and awareness-raising campaigns on the benefits of a green hydrogen-empowered economy.

## SYSTEM MODELLING

The proposed hybrid model of the GEN/PV/WT/battery system for green hydrogen production has been presented by a block diagram shown in Fig. 3. This diagram provides a system integration overview consisting of the generator, electrolyzer, solar panels, wind turbines, and battery storage combination to generate and store green hydrogen in a self-sufficient manner.

*Figure 3. Schematic of proposed hybrid microgrid model for hydrogen production*

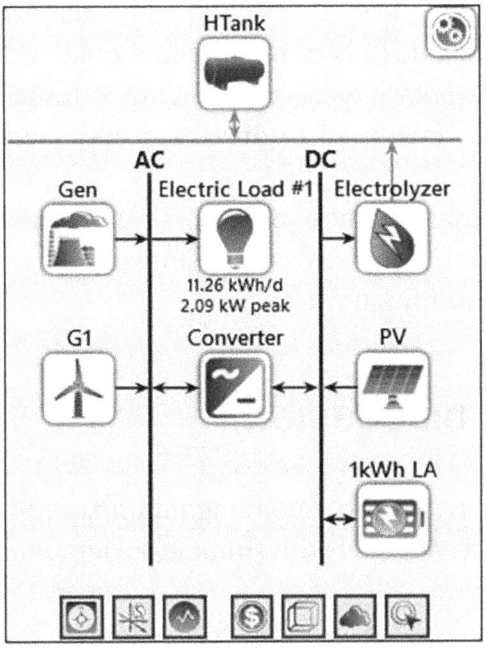


The proposed unified model for both electricity and green hydrogen production in a microgrid consists of the following major components:

a) *Renewable energy sources:* The RE sources include solar panels, wind turbines, or a combination of both. These sources generate electricity that is used to power the community and to produce hydrogen through the electrolysis process.
b) *Electrolyzer:* It is used to split water into hydrogen and oxygen using electricity from renewable energy sources. The hydrogen produced is stored in a tank and used to generate electricity through a fuel cell.
c) *Fuel cell:* It is used to generate electricity using the hydrogen produced by the electrolyzer. The electricity generated is used to power homes, businesses houses,and community facilities.
d) *Hydrogen storage tank:* It is used to store the hydrogen produced by the electrolyzer. The size of the tank depends on the hydrogen production capacity and the energy needs of the community.

e) *Control system:* It consists of sensors, controllers and software used to monitor and regulate the performance of the system. The control system ensures that the system operates efficiently and safely and that the electricity produced is of the required quality.
f) *Electricity distribution system:* It includes transformers, inverters, and distribution lines used to distribute the electricity generated to the community.
g) *Energy storage system:* The energy storage system includes batteries or other energy storage devices used to store excess electricity generated by the RE system for later use.

## RESULTS AND DISCUSSION

Energy generation and corresponding minimization of greenhouse gas emissions have been analyzed with simulation outcomes in the successive section:

*(i). Renewable penetration and excess electricity utilization:* Renewable penetration refers to the amount of total energy demand that is met by renewable sources such as solar, wind, electrolyzer, and generator. This graph illustrates the percentage of energy demand fulfilled by various RE sources such as solar, wind turbine, etc. as presented in Table 2 and Fig. 4.

*Figure 4. Characteristics of renewable penetration and load compliance*

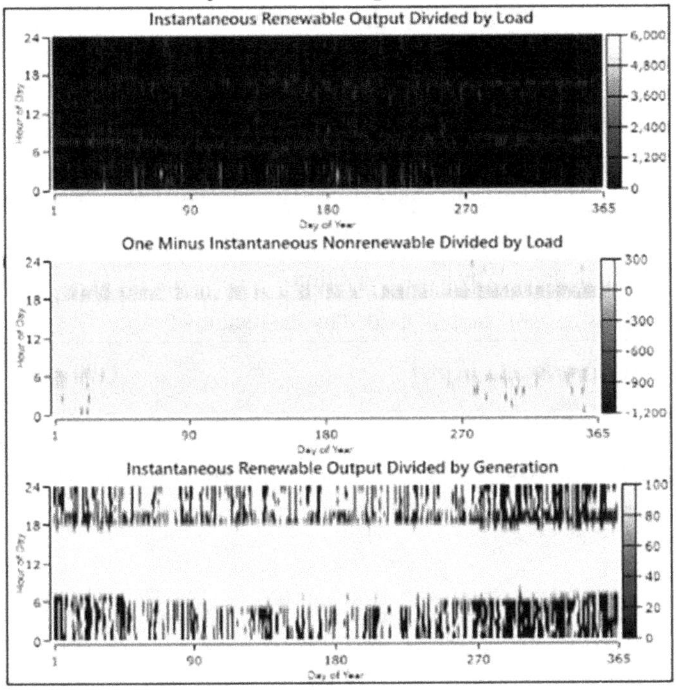


*Table 2. Results for RE generation and consumption*

| Sl. | Components | kWh/year | % |
|---|---|---|---|
| 1 | Generic flat PV plate | 27719 | 86.6 |
| 2 | Generic WT | 3719 | 11.6 |
| 3 | AC primary load | 4109 | 15.6 |
| 4 | DC primary load | 0 | 0 |
| 5 | Deferrable load | 0 | 0 |
| 6 | Excess electricity | 5393 | 16.8 |
| 8 | Unmet electric load | 0 | 0 |
| 9 | Capacity shortage | 0 | 0 |
| 10 | Renewable fraction | - | 86.0 |
| 11 | Maximum renewable penetration | 5945 | - |
| 12 | Nominal renewable capacity divided by total nominal capacity | - | 89.9 |

continued on following page

*Table 2. Continued*

| Sl. | Components | kWh/year | % |
|---|---|---|---|
| 13 | Usable renewable capacity divided by total capacity | - | 86.0 |
| 14 | Total renewable production divided by load | - | 120 |
| 15 | Total renewable production divided by generation | - | 98.2 |
| 16 | One minus total nonrenewable production divided by the load | - | 97.8 |
| 17 | Renewable output divided by load (HOMER standard) | - | 5945 |
| 18 | Renewable output divided by total generation | - | 100 |
| 19 | One minus nonrenewable output divided by the total load | - | 100 |

*Figure 5. Characteristics of solar and wind power generation annually*

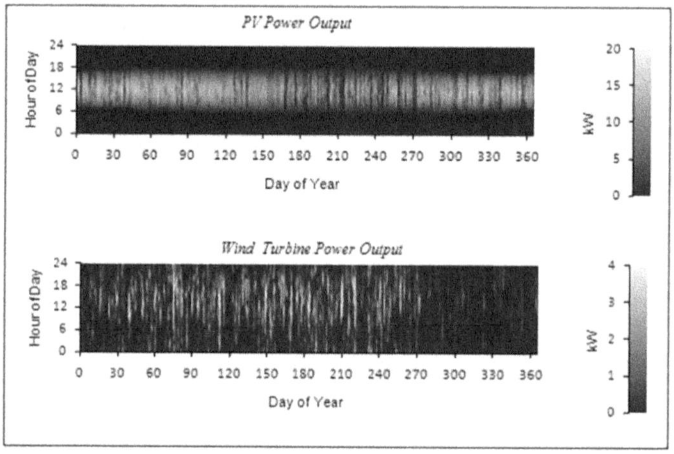


*(ii). Techno-economic analysis for the unified model:* The proposed model has multiple components for system integration and their techno-economic analysis has been shown in successive studies:

*(a). Solar power generation:* Solar power is an abundant RE source used in environment-friendly energy production. Solar PV converts sunlight into electricity, which can then be used to power an electrolyzer to produce hydrogen from water. The electricity generated by solar panels is intermittent and depends on the availability of sunlight. Therefore, to ensure reliable hydrogen production, the solar PV system must be appropriately sized and integrated with an energy

storage system such as batteries or hydrogen storage. The PV panel's capital and replacement costs have been estimated as 2500$/kW, while O&M costs are assumed to be 10$/year. The detailed specifications of the PV panels along with key results are presented in Table 3. The PV panel has a derating factor of 80% and an efficiency of 16%. The capacity of the PV panel is optimized using HOMER optimizer tools. The characteristic for PV generation is presented along with other performance metrics in Fig. 5.

*(b). Wind power generation:* Wind turbines are another key RE component used in green energy production. It generates electricity by harnessing the kinetic energy of wind, then it is used to power an electrolyzer to produce hydrogen from water. Like sunlight, a wind resource is intermittent and depends on the availability of potential windzones. To ensure reliable hydrogen production, the wind turbine system should be appropriately sized and integrated with an energy storage system such as battery backup or hydrogen storage. Overall, WTs are an important component in a microgrid-based green hydrogen unit. It can be used to coproduce hydrogen and generate electricity from the microgrid while ensuring reliable and cost-effective operation. The capital, replacement and O&M costs of the WT are assumed to be $7000, $7000 and 70 $/kW/year, respectively. It has a lifetime of 20 years with a hub height of 17m. The output of the WTis optimized using HOMER modeling tools. The performance characteristics and power output recorded results have been illustrated in Fig. 5 and Table 3.

*(c). Battery energy storage:* These storage systems are commonly used in microgrids. The batteries store excess energy generated during periods of high generation and it during periods of peak demand with low generation. Here, the LA batteries in the hybrid RE system are used as energy storage devices to preserve excess generated energy. The observations for state of charge along with other performance metrics of the LAB storage system have been shown in Fig. 6 and Table 3.

*Figure 6. Characteristics for battery energy charging cycle*

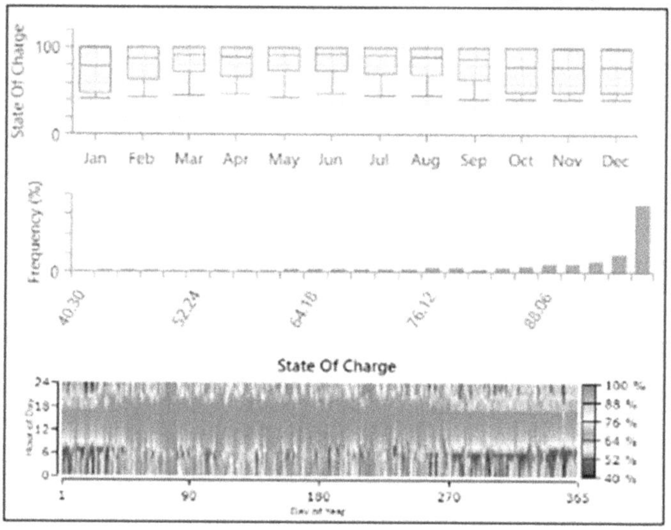


*(d). Power conversion system:* In a microgrid, the converters play an important role in the integration of RE sources and energy storage systems. These electrical devices convert the voltage, frequency, and current of the electricity generated by the RE sources and stored in the energy storage system to a compatible level for use by the microgrid.

*Figure 7. Characteristics of power conversion system*

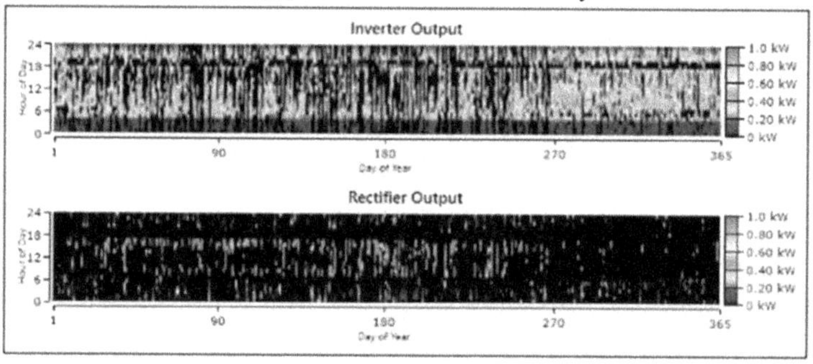

The capital, replacement and O&M costs of the converter are assumed to be \$300, \$300 and 00 \$/kW/year, respectively. In this study, the incorporated converter has a lifetime of 15 years with the 95% efficiency. The capacity of the converter is optimized using HOMER software. Its performance metrics for power conversion while working as an inverter and rectifier is presented in Fig. 7 and Table 3.

*(e). Hydrogen production with electrolyzer:* The electrolyzer is a critical component for green hydrogen production in a microgrid. It is an electrical device that splits water into hydrogen and oxygen using electricity. The hydrogen produced by the electrolyzer can be used as a fuel for various applications, such as power generation, transportation, and industrial processes. In this study, the electrolyzer has taken 15 year's lifetimes along with 80% efficiency. The recorded input power characteristics of the electrolyzer along with other performance metrics are presented in Fig. 8 and Table 3.

*Figure 8. Characteristics for hydrogen production of electrolyzer*

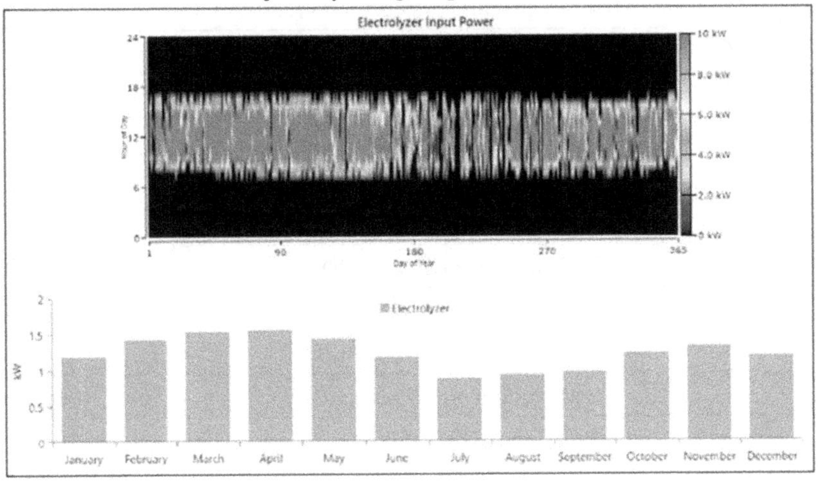

*(f). Hydrogen storage tank performance:* The sizing of the electrolyzer depends on the hydrogen production rate required by the microgrid. The electrolyzer must also be optimized to match the hydrogen production rate with the demand for hydrogen. The electrolyzer can be powered by RE sources such as solar PV and wind turbines to ensure that the hydrogen produced is green. In this study, the capacity optimization for hydrogen production achieved is 100kg, and relative to tank size is 35%.Its performance characteristics suggested that the total production of electrolyzer in hydrogen tank in kg/year with LCOH of 2.32 $/kg as shown in Fig. 8 and Table 3.

*Table 3. Specification of the integral components of the proposed hybrid energy system*

| Sl. | Components | Parameter | Value |
|---|---|---|---|
| 1 | PV module | PV slope (°) | 36 |
| | | Azimuth (°) | 0 |
| | | Derating factor (%) | 80 |
| | | Temperature effects on power (%/° C) | -0.38 |
| | | Nominal operating cell temperature (° C) | 45 |
| | | Efficiency of PV arrays at STC (%) | 16 |
| | | Lifetime (years) | 25 |
| | | Hours of operation (hours/year) | 4383 |
| | | Rated capacity (kW) | 16.4 |
| | | Mean power output (kW) | 3.16 |
| | | Mean energy output (kWh/day) | 75.9 |
| | | Minimum power output (kW) | 0 |
| | | Maximum power output (kW) | 16.2 |
| | | Capacity factor (%) | 19.2 |
| | | Capital cost ($) | 411.17 |
| | | Specific yield (kWh/kW) | 1,685 |
| | | PV penetration (%) | 675 |
| | | Total production (kW) | 27,719 |

continued on following page

*Table 3. Continued*

| Sl. | Components | Parameter | Value |
|---|---|---|---|
| | | Maintenance cost ($/year) | 1.64 |
| | | LCOE ($/kWh) | 0.00121 |
| 2 | Wind turbine | Rated power capacity (kW) | 4 |
| | | Capital cost ($/unit) | 14,00 |
| | | Hours of operation (hours/year) | 6316 |
| | | O&M cost ($/year/unit) | 14 |
| | | Lifetime (years) | 20 |
| | | Wind power generation cost($/year/unit) | 7000 |
| | | Hub height (m) | 17 |
| | | Wind turbine total production (kWh/year) | 3719 |
| | | Mean output (kW) | 0.425 |
| | | Capacity factor (%) | 10.6 |
| | | Minimum power output (kW) | 0 |
| | | Maximum power output (kW) | 4 |
| | | Wind penetration (%) | 90.5 |
| | | Quantity | 4 |
| | | LCOE ($/kWh) | 0.0369 |
| 3 | Battery storage | Energy in (kWh/year) | 1046 |
| | | Energy out (kWh/year) | 839 |
| | | Nominal capacity (kWh) | 7.01 |
| | | Usable nominal capacity (kWh) | 4.20 |
| | | Cyclic performance efficiency (%) | 80 |
| | | Autonomy duration (hour) | 8.96 |
| | | Average energy cost ($/kWh) | 0 |
| | | Storage depletion (kWh/year) | 2.58 |
| | | Losses (kWh/year) | 210 |
| | | Annual throughput (kWh/year) | 938 |
| | | Maintenance cost ($/year) | 70.0 |
| | | Storage wear cost ($/kWh) | 0.419 |
| | | Capital cost ($) | 2100 |
| | | Lifetime throughput (kWh) | 5600 |
| | | Strings in parallel | 7 |

continued on following page

*Table 3. Continued*

| Sl. | Components | Parameter | Value |
|---|---|---|---|
| | | String size | 1 |
| | | Bus voltage | 12 |
| | | Quantity | 7 |
| | | Expected life (year) | 5.97 |
| 4 | Electrolyzer | Minimum load ratio (%) | 35 |
| | | Lifetime (years) | 15 |
| | | Efficiency (%) | 80 |
| | | Rated power capacity (kW) | 10 |
| | | Total input energy (kWh/year) | 22173 |
| | | Power input to output variation (kW) | 0 - 2.53 |
| | | Minimum hydrogen output (kg/hour) | 0 |
| | | Maximum power output (kW) | 10 |
| | | Maximum hydrogen output (kg/hour) | 0.203 |
| | | Mean hydrogen output (kg/hour) | 0.0513 |
| | | Total hydrogen production (kg/year) | 450 |
| | | Hours of operation (hour/year) | 2806 |
| | | Capacity factor (%) | 25.3 |
| | | Specific consumption (kWh/kg) | 49.3 |
| 5 | Power converters | Converter efficiency (%) | 95 |
| | | Converter lifetime (years) | 15 |
| | | Converter capital cost ($/kW) | 300 |
| | | Inverter capacity (kW) | 0.972 |
| | | Rectifier capacity (kW) | 0.972 |
| | | Inverter mean output (kW) | 0.222 |
| | | Rectifier mean output (kW) | 0.0598 |
| | | Inverter minimum output (kW) | 0 |
| | | Rectifier minimum output (kW) | 0 |
| | | Inverter maximum output (kW) | 0.883 |
| | | Rectifier maximum output (kW) | 0.972 |
| | | Inverter hours of operation (hours/year) | 5872 |
| | | Rectifier hours of operation (hours/year) | 1819 |
| | | Inverter energy out (kWh/year) | 1945 |

continued on following page

*Table 3. Continued*

| Sl. | Components | Parameter | Value |
|---|---|---|---|
| | | Rectifier energy out (kWh/year) | 524 |
| | | Inverter energy in (kWh/year) | 2047 |
| | | Rectifier energy in (kWh/year) | 524 |
| | | Inverter losses (kWh/year) | 102 |
| | | Rectifier losses (kWh/year) | 27.6 |
| | | Inverter capacity factor (%) | 22.8 |
| | | Rectifier capacity factor (%) | 6.16 |
| | | Rated power capacity (kWh) | 6 |
| | | Annual throughput (kWh/year) | 723 |
| | | Operation cost ($/year) | 300 |
| | | Maintenance cost ($/year) | 60 |
| | | Autonomy duration (hour) | 7.68 |
| | | Expected life (year) | 6.64 |
| | | Capital costs ($) | 1800 |
| | | Losses (kWh/year) | 162 |
| | | Efficiency (%) | 60 |
| 7 | Hydrogen tank | Capital cost ($/kg) | 1362 |
| | | O&M cost ($/kg) | 68 |
| | | Initial tank level (%) | 35 |
| | | Lifetime (year) | 15 |
| | | Hydrogen storage capacity (kg) | 100 |
| | | Content at beginning of year (kg) | 50 |
| | | Content at end of year (kg) | 100 |
| | | Tank autonomy (hour) | 7106 |
| | | Energy storage capacity (kWh) | 3333 |
| | | LCOH ($/kg) | 2.32 |
| | | Available hydrogen (kg/year) | 450 |

(iii). Optimization results and emission reduction with green hydrogen production: The production optimization of green electricity and hydrogen, supports low-carbon environmental protection and further supports the clean transformation of energy structure. The results of the present study are given below:

(a). Optimization Results for techno-economic objective

The optimization results provide valuable insights into the performance and efficiency of a RE system, enabling system designers and operators to identify opportunities for improvement and maximize the energy output of the system. Corresponding optimization results for technical and economical solutions have been presented in Table 4(a) - 4(b), respectively.

*Table 4. Optimization results for technical aspects of the system*

| | | | | | | PV (kW) | WT (kW) | LAB (kWh) | EL (kW) | HT (kg) | PC (kW) |
|---|---|---|---|---|---|---|---|---|---|---|---|
| | | | | | - | 16.4 | 2.30 | 7 | 10.0 | - | 0.972 |
| | | | | | | 16.4 | 2.30 | 7 | 10.0 | 100 | 0.972 |
| | | | | | - | 14.6 | - | 14 | 10.0 | - | 1.91 |
| | | | | | | 14.6 | - | 14 | 10.0 | 100 | 1.91 |
| | - | | | | - | 20.7 | 2.30 | 9 | 10.0 | - | 1.03 |
| | - | | | | | 20.7 | 2.30 | 9 | 10.0 | 100 | 1.03 |

| | | | | | | Dispatch | NPC ($) | COE ($) | Operating cost ($/year) | Initial capital ($) | RF (%) |
|---|---|---|---|---|---|---|---|---|---|---|---|
| | | | | | - | LF | 13510 | 0.254 | 630.97 | 5353 | 86.0 |
| | | | | | | LF | 13510 | 0.254 | 630.97 | 5353 | 86.0 |
| | | | | | - | CC | 14328 | 0.270 | 602.54 | 6539 | 100 |
| | | | | | | CC | 14328 | 0.270 | 602.54 | 6539 | 100 |
| | - | | | | - | LF | 15748 | 0.296 | 856.43 | 4676 | 81.7 |
| | - | | | | | LF | 15748 | 0.296 | 856.43 | 4676 | 81.7 |

(b). Minimization of emission

The proposed system with RE sources, such as solar, wind, electrolyzer can significantly reduce greenhouse gas emissions and other harmful pollutants, making it a viable solution for mitigating climate change and achieving a more sustainable energy future. In Table 5, the optimization result for emission reduction has been shown.

*Table 5. Results of emission reduction*

| Sl | Pollutant | Quantity |
|---|---|---|
| 1 | Carbon dioxide (kg/year) | 560 |
| 2 | Carbon monoxide (kg/year) | 3.53 |
| 3 | Unburned hydrocarbons (kg/year) | 0.154 |
| 4 | Particulate matter (kg/year) | 0.0214 |
| 5 | Sulfur dioxide (kg/year) | 1.37 |
| 6 | Nitrogen oxides (kg/year) | 3.31 |

*Figure 9. Characteristics for cash flow and cost summary of the proposed system*

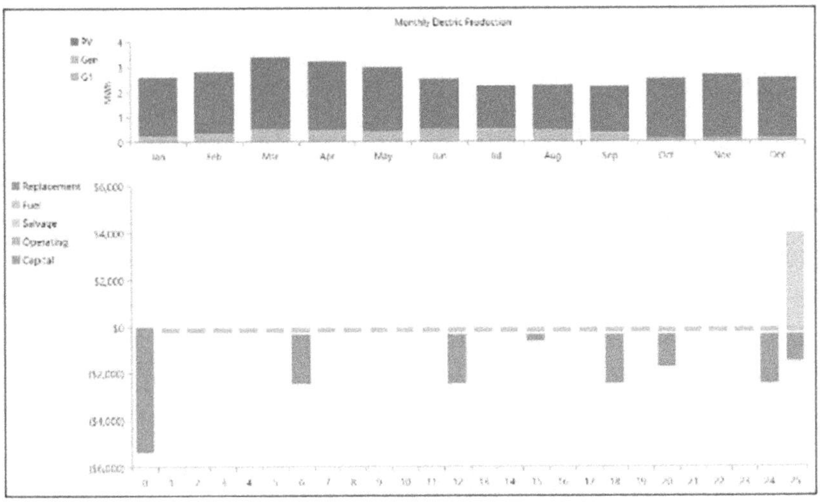


(c). Economic analysis

This cash flow characteristic illustrates the value of energy demand fulfilled by various RE sources such as solar, wind, etc. It provides correlation among cash inflows and outflows, replacement cost, salvage value, operating cost, capital over the lifetime of the proposed work, it help stakeholders to evaluate the financial feasibility and profitability of investments in RE sector. A summary of costs associated with different RE sources provides an overview of the expenses involved in generating, storing, and distributing energy from sources such as generators, electrolyzer, solar, wind, etc. In this study, the key parameters for financial feasibility and is shown in Table 6 and Fig. 9.

*Table 6. Financial evaluation rates for the proposed system*

| Financial particulars | Rate |
|---|---|
| Discount rate (%) | 8.00 |
| Inflation rate (%) | 2.00 |
| Annual capacity storage (%) | 0.00 |
| Project lifetime (years) | 25.00 |

## CONCLUSION

The analysis of the Chintamanpur region in Bihar suggests significant potential for RE generation, especially through solar and wind sources. The annual average solar radiation is 5.22 kWh/m$^2$/day, the average wind speed is 4.33 m/s and the annual temperature resource is 26.14 °C, which makes it a favorable location for developing hybrid RE systems. Based on the HOMER simulation results, the most efficient system with the lowest energy cost per kWh is the given scenario with a NPC of approximately $13,509.64 and an energy cost of $0.2543 kWh. This scenario includes a hybrid system with solar panels, wind turbines, generators, batteries, hydrogen tanks and electrolyzer, which can meet the annual load and generate excess electricity. The emission of carbon dioxide is 560 kg/year with a simple payback period of around 1.4 years. Overall, the findings suggest that green hydrogen unit in a microgrid can play a crucial role in the transition to a more sustainable energy system.

## REFERENCES


Aayog, N. I. T. I. (2022). Green Hydrogen Is Critical to India's Economic Development and Net-Zero Ambitions: Report, https://www.pib.gov.in/PressReleasePage.aspx?PRID=1838104

Akarsu, B., & Genç, M. S. (2022). Optimization of electricity and hydrogen production with hybrid renewable energy systems. *Fuel*, *324*, 124465. DOI: 10.1016/j.fuel.2022.124465

Akhtari, M. R., & Baneshi, M. (2019). Techno-economic assessment and optimization of a hybrid renewable co-supply of electricity, heat, and hydrogen system to enhance performance by recovering excess electricity for a large energy consumer. *Energy Conversion and Management*, *188*, 131–141. DOI: 10.1016/j.enconman.2019.03.067

Asian Development Bank. (2023), Hydrogen market in India. chrome-extension://efaidnbmnnnibpcajpcglclefindmkaj/https://teriin.org/sites/default/files/files/India-Scenarios-Sketch.pdf

Azad, A., & Shateri, H. (2023). Design and optimization of an entirely hybrid renewable energy system (WT/PV/BW/HS/TES/EVPL) to supply electrical and thermal loads with considering uncertainties in generation and consumption. *Applied Energy*, *336*, 120782. DOI: 10.1016/j.apenergy.2023.120782

Bachmann, O. (2023). Large scale low-cost green hydrogen production using thermal energy storage and polymer electrolyte membrane electrolysis systems. *IET Renewable Power Generation*, *17*(4), 775–793. DOI: 10.1049/rpg2.12633

Bryan, J., Meek, A., Dana, S., Sakir, M. S. I., & Wang, H. (2023). Modeling and design optimization of carbon-free hybrid energy systems with thermal and hydrogen storage. *International Journal of Hydrogen Energy*, *48*(99), 39097–39111. DOI: 10.1016/j.ijhydene.2023.03.135

Di Micco, S., Romano, F., Jannelli, E., Perna, A., & Minutillo, M. (2023). Techno-economic analysis of a multi-energy system for the co-production of green hydrogen, renewable electricity and heat. *International Journal of Hydrogen Energy, 48*(81), 31457–31467. DOI: 10.1016/j.ijhydene.2023.04.269

Eisapour, A. H., Jafarpur, K., & Farjah, E. (2021). Feasibility study of a smart hybrid renewable energy system to supply the electricity and heat demand of Eram Campus, Shiraz University; simulation, optimization, and sensitivity analysis. *Energy Conversion and Management, 248*, 114779. DOI: 10.1016/j.enconman.2021.114779

Fan, G., Liu, Z., Liu, X., Shi, Y., Wu, D., Guo, J., Zhang, S., Yang, X., & Zhang, Y. (2022). Two-layer collaborative optimization for a renewable energy system combining electricity storage, hydrogen storage, and heat storage. *Energy, 259*, 125047. DOI: 10.1016/j.energy.2022.125047

Hai, T., El-Shafay, A. S., Alizadeh, A. A., Dhahad, H. A., Chauhan, B. S., Almojil, S. F., Almohana, A. I., & Alali, A. F. (2023). Comparison analysis of hydrogen addition into both anode and afterburner of fuel cell incorporated with hybrid renewable energy driven SOFC: An application of techno-environmental horizon and multi-objective optimization. *International Journal of Hydrogen Energy.*

Han, F., Zeng, J., Lin, J., & Gao, C. (2023). Multi-stage distributionally robust optimization for hybrid energy storage in regional integrated energy system considering robustness and nonanticipativity. *Energy, 277*, 127729. DOI: 10.1016/j.energy.2023.127729

HassanzadehFard, H., Tooryan, F., & Dargahi, V. (2023). Standalone hybrid system energy management optimization for remote village considering methane production from livestock manure. *International Journal of Hydrogen Energy, 48*(29), 10778–10796. DOI: 10.1016/j.ijhydene.2022.12.085

He, J., Wu, Y., Wu, M., Xu, M., & Liu, F. (2022). Two-stage configuration optimization of a novel standalone renewable integrated energy system coupled with hydrogen refueling. *Energy Conversion and Management, 251*, 114953. DOI: 10.1016/j.enconman.2021.114953

He, J., Wu, Y., Yong, X., Tan, Q., & Liu, F. (2022). Bi-level optimization of a near-zero-emission integrated energy system considering electricity-hydrogen-gas nexus: A two-stage framework aiming at economic and environmental benefits. *Energy Conversion and Management*, *274*, 116434. DOI: 10.1016/j.enconman.2022.116434

Huang, J., Balcombe, P., & Feng, Z. (2023). Technical and economic analysis of different colours of producing hydrogen in China. *Fuel*, *337*, 127227. DOI: 10.1016/j.fuel.2022.127227

Huangfu, Y., Tian, C., Zhuo, S., Xu, L., Li, P., Quan, S., Zhang, Y., & Ma, R. (2023). An optimal energy management strategy with subsection bi-objective optimization dynamic programming for photovoltaic/battery/hydrogen hybrid energy system. *International Journal of Hydrogen Energy*, *48*(8), 3154–3170. DOI: 10.1016/j.ijhydene.2022.10.133

Izadi, A., Shahafve, M., & Ahmadi, P. (2022). Neural network genetic algorithm optimization of a transient hybrid renewable energy system with solar/wind and hydrogen storage system for zero energy buildings at various climate conditions. *Energy Conversion and Management*, *260*, 115593. DOI: 10.1016/j.enconman.2022.115593

Izadi, A., Shahafve, M., Ahmadi, P., & Hanafizadeh, P. (2023). Design, and optimization of COVID-19 hospital wards to produce Oxygen and electricity through solar PV panels with hydrogen storage systems by neural network-genetic algorithm. *Energy*, *263*, 125578. DOI: 10.1016/j.energy.2022.125578 PMID: 36189102

Jia, K., Liu, C., Li, S., & Jiang, D. (2023). Modeling and optimization of a hybrid renewable energy system integrated with gas turbine and energy storage. *Energy Conversion and Management*, *279*, 116763. DOI: 10.1016/j.enconman.2023.116763

Jurado, F., Mezrhab, A., Moussaoui, M. A., & Vera, D. (2023). Cost and size optimization of hybrid solar and hydrogen subsystem using HomerPro software. *International Journal of Hydrogen Energy*.

Khan, T., Yu, M., & Waseem, M. (2022). Review on recent optimization strategies for hybrid renewable energy system with hydrogen technologies: State of the art, trends and future directions. *International Journal of Hydrogen Energy, 47*(60), 25155–25201. DOI: 10.1016/j.ijhydene.2022.05.263

Kilic, M., & Altun, A. F. (2022). Dynamic modelling and multi-objective optimization of off-grid hybrid energy systems by using battery or hydrogen storage for different climates. *International Journal of Hydrogen Energy*.

Liu, J., Zhou, Y., Yang, H., & Wu, H. (2022). Net-zero energy management and optimization of commercial building sectors with hybrid renewable energy systems integrated with energy storage of pumped hydro and hydrogen taxis. *Applied Energy, 321*, 119312. DOI: 10.1016/j.apenergy.2022.119312

Luo, L., Cristofari, C., & Levrey, S. (2023). Cogeneration: Another way to increase energy efficiency of hybrid renewable energy hydrogen chain–A review of systems operating in cogeneration and of the energy efficiency assessment through exergy analysis. *Journal of Energy Storage, 66*, 107433. DOI: 10.1016/j.est.2023.107433

Maestre, V. M., Ortiz, A., & Ortiz, I. (2022). Transition to a low-carbon building stock. Techno-economic and spatial optimization of renewables-hydrogen strategies in Spain. *Journal of Energy Storage, 56*, 105889. DOI: 10.1016/j.est.2022.105889

Mehrenjani, J. R., Gharehghani, A., Nasrabadi, A. M., & Moghimi, M. (2022). Design, modeling and optimization of a renewable-based system for power generation and hydrogen production. *International Journal of Hydrogen Energy, 47*(31), 14225–14242. DOI: 10.1016/j.ijhydene.2022.02.148

Ministry of Environment. Forest and Climate Change (2023). Net zero emissions target. https://pib.gov.in/PressReleaseIframePage.aspx?PRID=1945472#:~:text=India%2C%20at%20the%2026th%20session,achieve%20net%20zero%20by%202070

Qiu, R., Zhang, H., Wang, G., Liang, Y., & Yan, J. (2023). Green hydrogen-based energy storage service via power-to-gas technologies integrated with multi-energy microgrid. *Applied Energy, 350*, 121716. DOI: 10.1016/j.apenergy.2023.121716

Shahverdian, M. H., Sedayevatan, S., Hosseini, M., Sohani, A., Javadijam, R., & Sayyaadi, H. (2023). Multi-objective technoeconomic optimization of an off-grid solar-ground-source driven cycle with hydrogen storage for power and fresh water production. *International Journal of Hydrogen Energy*, *48*(52), 19772–19791. DOI: 10.1016/j.ijhydene.2023.02.062

S&P Global Commodities Insights. (2023). Global hydrogen demand to more than triple by 2050, driven by power generation, aviation and industry: analyst. https://www.hydrogeninsight.com/production/global-hydrogen-demand-to-more-than-triple-by-2050-driven-by-power-generation-aviation-and-industry-analyst/2-1-1394924

The Energy and Resources Institute. (2023). India transforming to a net-zero emissions energy system - A call to action to 2030. chrome-extension://efaidnbmnnnibpcajpcglclefindmkaj/https://teriin.org/sites/default/files/files/India-Scenarios-Sketch.pdf

The Indian Express. (2023). Green Hydrogen push will need to counter challenges. https://indianexpress.com/article/opinion/editorials/green-hydrogen-push-will-need-to-counter-challenges-8957673/

UN DESA. (2023). UN Department of Economic and Social Affairs (DESA) Policy Brief No. 153: India overtakes China as the world's most populous country, https://www.un.org/development/desa/dpad/publication/un-desa-policy-brief-no-153-india-overtakes-china-as-the-worlds-most-populous-country/

Wang, R., & Zhang, R. (2023). Techno-economic analysis and optimization of hybrid energy systems based on hydrogen storage for sustainable energy utilization by a biological-inspired optimization algorithm. *Journal of Energy Storage*, *66*, 107469. DOI: 10.1016/j.est.2023.107469

Wu, S., Lin, J., Li, J., & Song, Y. (2022). Electricity allocation strategy for on-site hydrogen refueling stations in the forward and spot markets. *IET Renewable Power Generation*, *16*(12), 2532–2546. DOI: 10.1049/rpg2.12406

Yan, R., Wang, J., Huo, S., Qin, Y., Zhang, J., Tang, S., Wang, Y., Liu, Y., & Zhou, L. (2023). Flexibility improvement and stochastic multi-scenario hybrid optimization for an integrated energy system with high-proportion renewable energy. *Energy, 263*, 125779. DOI: 10.1016/j.energy.2022.125779

Yang, G., Zhang, H., Wang, W., Liu, B., Lyu, C., & Yang, D. (2023). Capacity optimization and economic analysis of PV–hydrogen hybrid systems with physical solar power curve modeling. *Energy Conversion and Management, 288*, 117128. DOI: 10.1016/j.enconman.2023.117128

Zhang, K., Zhou, B., Chung, C. Y., Bu, S., Wang, Q., & Voropai, N. (2022). A coordinated multi-energy trading framework for strategic hydrogen provider in electricity and hydrogen markets. *IEEE Transactions on Smart Grid, 14*(2), 1403–1417. DOI: 10.1109/TSG.2022.3154611

Zhang, K., Zhou, B., Or, S. W., Li, C., Chung, C. Y., & Voropai, N. (2021). Optimal coordinated control of multi-renewable-to-hydrogen production system for hydrogen fueling stations. *IEEE Transactions on Industry Applications, 58*(2), 2728–2739. DOI: 10.1109/TIA.2021.3093841

Zhang, L., Qiu, Y., Chen, Y., & Hoang, A. T. (2023). Multi-Objective Particle Swarm Optimization applied to a solar-geothermal system for electricity and hydrogen production; Utilization of zeotropic mixtures for performance improvement. *Process Safety and Environmental Protection, 175*, 814–833. DOI: 10.1016/j.psep.2023.05.082

Zhang, Y., Hua, Q. S., Sun, L., & Liu, Q. (2020). Life cycle optimization of renewable energy systems configuration with hybrid battery/hydrogen storage: A comparative study. *Journal of Energy Storage, 30*, 101470. DOI: 10.1016/j.est.2020.101470

Zhao, P., Gu, C., Hu, Z., Xie, D., Hernando-Gil, I., & Shen, Y. (2020). Distributionally robust hydrogen optimization with ensured security and multi-energy couplings. *IEEE Transactions on Power Systems, 36*(1), 504–513. DOI: 10.1109/TPWRS.2020.3005991

Zheng, L., Wang, J., Yu, Y., Li, G., Zhou, M., Xia, Q., & Xu, G. (2022). On the consistency of renewable-to-hydrogen pricing. *CSEE Journal of Power and Energy Systems, 8*(2), 392–402.

Zheng, Y., You, S., Bindner, H. W., & Münster, M. (2022). Incorporating optimal operation strategies into investment planning for wind/electrolyzer system. *CSEE Journal of Power and Energy Systems*, *8*(2), 347–359.

Zhou, X., Zhao, Q., Zhang, Y., & Sun, L. (2021). Integrated energy production unit: An innovative concept and design for energy transition toward low-carbon development. *CSEE Journal of Power and Energy Systems*, *7*(6), 1133–1139.

## KEY TERMS AND DEFINITIONS

**BT:** Battery.

**CI:** Clearness Index.

**DG:** Diesel Generator.

**DR:** Daily Radiation (kWh/m2/day).

**EE:** Energy Excess.

**EL:** Electrolyzer.

**EP:** Energy Production.

**ES:** Energy Storage.

**EU:** Energy Utilization.

**FCV:** Fuel Cell Vehicles.

**FLH:** Full Load Hours.

**GHI:** Global Horizontal Irradiance.

**HDI:** Human Development Index.

**HESS:** Hybrid Energy Storage System.

**HFS:** Hydrogen Fueling Station.

**HP:** Hybrid Renewable to H2 Provider.

**HRES:** Hybrid Renewable Energy System.

**HSS:** Hydrogen Storage Systems.

**HT:** Hydrogen Tank.

**HY:** Hydrogen.

**HYP:** Hydrogen Production.

**IEGS:** Integrated Electricity AndGas System.

**IEPU:** Integrated Energy Production Unit.

**LA:** Lead Acid.

**LAB:** LA Battery.

**LCOE:** Levelized Cost of Energy.

**LCOH:** Levelized Cost of Hydrogen.

**LMP:** Locational Marginal Price.

**LMHP:** Locational Marginal Hydrogen Price.

**Nw:** Average Wind Speed (m/s).

**NPC:** Net Present Cost.

**OHRS:** On Site Hydrogen Refueling Station.

**O&M:** Operation and Maintenance Cost.

**T:** Daily temperature (oC).

**PC:** Power Converter.

**PER:** Pollutant Emission Reduction.

**P2G:** Power-to-Gas.

**PV:** Photovoltaic.

**PVF:** Present Value Factor.

**RE:** Renewable Energy.

**RF:** Renewable Fraction.

**WT:** Wind Turbine.

# Conclusion

This book introduces hydrogen, which is used as a basic resource in many industries, and hydrogen production is expected to contribute significantly to the future development of the hydrogen energy economy. The book "Developing Advanced Technologies for Hydrogen Storage and Production" focuses on hydrogen-based energy systems. Critical reviews that provide insight into the topic from recognized global specialists are considered. Studies may include, but are not limited to, topics aimed at advancing energy and environmental sustainability through the development of improved hydrogen production technologies, including electrolysis, thermochemical, electrochemical, photobiological, and photochemical methods. In addition, membrane separation, pressure swing adsorption, and cryogenic distillation can produce hydrogen-rich gases instead of pure hydrogen. This book covers renewable energy, nanotechnology, power generation, boron, military applications, health, food, environmental challenges, and aerospace.

The Initiative for Developing Advanced Technologies for Hydrogen Storage and Production book uses a novel methodology to achieve its goals. This book discusses the advancement of advanced technologies for hydrogen storage and production. This document will facilitate the exchange of ideas, best practices, and advances in fundamental and industrial hydrogen research and engineering. The study of hydrogen and its control covers a wide range of science.

This book examines theoretical and practical engineering science and technology to illuminate advances in hydrogen technology. It aims to provide scientists, engineers, technologists, industry experts and students with relevant and important data on hydrogen production technology research and regulation.

The target audience for The Developing Advanced Technologies for Hydrogen Storage and Production will consist of scientists, engineers, technologists, and industry professionals involved in many hydrogen-related fields as follows:

Section 1 focuses on "Improving Hydrogen Energy". Hydrogen energy systems, in addition to being a green clean energy source and high efficiency, have started to gain importance with new technological developments. In addition, we can also use hydrogen technology to analyze all discipline and field for the future. The development of hydrogen energy is the subject of investigation in Section 1, which is divided into three separate chapters. Challenges and barriers to the implementation of hydrogen energy are discussed in Chapter 1. Chapter 2 provides an overview of the path to a hydrogen economy. It is the development of hydrogen energy that is highlighted in this section. The developments and difficulties in the production of green hydrogen are outlined in Chapter 3.

Section 2 deals with "Applications of hydrogen technology in the automotive and industrial sectors". Applications of hydrogen technology in the automotive and industrial sectors are discussed in Section 2. Chapter 4 discusses green hydrogen for cement production. This chapter focuses on a decarbonization pathway that uses hybrid renewable energy sources. Finally, hydrogen technology can be rapidly implemented in a variety of industries, including the automotive and industrial sectors, as well as other areas. The use of plasmatron technology for hydrogen in fuel preparation and exhaust pollution reduction is described in Chapter 5.

Section 3 discusses "Renewable Energy Sources and the Green Carbon Cycle: The Future of Hydrogen. Section 3 discusses sustainable carbon cycle and renewable energy sources relevant to the future of hydrogen. As defined in Chapter 6, green computing is a complementary component in the production of hydrogen energy within an integrated renewable energy system. A discussion on the optimization of integrated power and green hydrogen generation using hybrid renewable energy sources is included in Chapter 7. In conclusion, we can advance future energy using hydrogen technology across all disciplines and fields.

# Compilation of References


Aayog, N. I. T. I. (2022). Green Hydrogen Is Critical to India's Economic Development and Net-Zero Ambitions: Report, https://www.pib.gov.in/PressReleasePage.aspx?PRID=1838104

Abhyankar, N., Mohanty, P., Deorah, S., Karali, N., Paliwal, U., Kersey, J. & Phadke, A. (2023). India's path towards energy independence and a clean future: Harnessing india's renewable edge for cost-effective energy independence by 2047. *2*, 36(5), p.107273.

Abot, M. D. (2020). Factors influencing adoption of artificial insemination by smallholder livestock farmers in dryland production systems of Kenya *(Doctoral dissertation, University of Nairobi)*.

Acar, C. (2024). *Hydrogen storage materials*. Reference Module in Earth Systems and Environmental Sciences., DOI: 10.1016/B978-0-44-313219-3.00113-1

Agnolucci, P., & Mcdowall, W. (2013). Designing future hydrogen infrastructure : Insights from analysis at different spatial scales. *International Journal of Hydrogen Energy*, *38*(13), 5181–5191. DOI: 10.1016/j.ijhydene.2013.02.042

Ahmadi, P., Dincer, I., & Rosen, M. A. (2013). Energy and exergy analyses of hydrogen production via solar-boosted ocean thermal energy conversion and PEM electrolysis. *International Journal of Hydrogen Energy*, *38*(4), 1795–1805. DOI: 10.1016/j.ijhydene.2012.11.025

Akarsu, B., & Genç, M. S. (2022). Optimization of electricity and hydrogen production with hybrid renewable energy systems. *Fuel*, *324*, 124465. DOI: 10.1016/j.fuel.2022.124465

Akhtari, M. R., & Baneshi, M. (2019). Techno-economic assessment and optimization of a hybrid renewable co-supply of electricity, heat, and hydrogen system to enhance performance by recovering excess electricity for a large energy consumer. *Energy Conversion and Management*, *188*, 131–141. DOI: 10.1016/j.enconman.2019.03.067

Alharbi, A. A., Alqahtani, N. B., Alkhedhair, A. M., Alabduly, A. J., Almaleki, A. A., Almadih, M. H., Albishi, M. S., & Almayeef, A. A. (2022). Developed plasmatron design to enhance production of hydrogen in synthesis gas produced by a fuel reformer system. *Energies*, *2022*(15), 1071. DOI: 10.3390/en15031071

Ali, A., Khan, M. A., & Choi, H. (2024). Supervised Machine Learning-Based Prediction of Hydrogen Storage Classes Utilizing Dibenzyltoluene as an Organic Carrier. *Molecules (Basel, Switzerland)*, *29*(6), 1280. DOI: 10.3390/molecules29061280 PMID: 38542916

Al-Shetwi, A. Q., Atawi, I. E., Abuelrub, A., & Hannan, M. A. (2023). Techno-economic assessment and optimal design of hybrid power generation-based renewable energy systems. *Technology in Society*, *75*, 102352. DOI: 10.1016/j.techsoc.2023.102352

Arpit, S., Pai, S., & Choudhary, S. M. (2024). Should state-owned coal companies harness green hydrogen? https://www.swaniti.com/wp-content/uploads/2024/10/Green-hydrogen-final.pdf

Arpit, S., & Das, P. K. (2023). A state-of-the-art review of heat recovery steam generators and waste heat boilers. *Energy Efficiency*, *16*(8), 99. Advance online publication. DOI: 10.1007/s12053-023-10179-5

Arpit, S., Kumar, P., Das, P. K., & Dash, S. K. (2023). Application of exergy analysis in understanding the performance of a coal-fired steam power plant (120 MW) with single reheat and regenerative configuration. *Journal of Thermal Engineering*, *9*(2), 497–509. DOI: 10.18186/thermal.1285229

Asian Development Bank. (2023), Hydrogen market in India. chrome-extension://efaidnbmnnnibpcajpcglclefindmkaj/https://teriin.org/sites/default/files/files/India-Scenarios-Sketch.pdf

Assareh, E., Agarwal, N., Paul, M. C., Ahmadi, P., Ghodrat, M., & Lee, M. (2023). Investigation and development of a novel solar-biomass integrated energy system for clean electricity and liquid hydrogen production. *Thermal Science and Engineering Progress*, *42*, 101925. DOI: 10.1016/j.tsep.2023.101925

Azad, A., & Shateri, H. (2023). Design and optimization of an entirely hybrid renewable energy system (WT/PV/BW/HS/TES/EVPL) to supply electrical and thermal loads with considering uncertainties in generation and consumption. *Applied Energy*, *336*, 120782. DOI: 10.1016/j.apenergy.2023.120782

Bachmann, O. (2023). Large scale low-cost green hydrogen production using thermal energy storage and polymer electrolyte membrane electrolysis systems. *IET Renewable Power Generation*, *17*(4), 775–793. DOI: 10.1049/rpg2.12633

Bag, S., Yadav, G., Dhamija, P., & Kataria, K. K. (2021). Key resources for industry 4.0 adoption and its effect on sustainable production and circular economy: An empirical study. *Journal of Cleaner Production*, *281*, 125233. DOI: 10.1016/j.jclepro.2020.125233

Ball, M., Weeda, M., Ball, M., & Weeda, M. (2015). The hydrogen economy — Vision or reality? *International Journal of Hydrogen Energy*, *40*(25), 7903–7919. DOI: 10.1016/j.ijhydene.2015.04.032

Baykara, S. Z. (2018). Hydrogen: A brief overview on its sources, production and environmental impact. *International Journal of Hydrogen Energy*, *43*(23), 10605–10614. DOI: 10.1016/j.ijhydene.2018.02.022

Bharathi, K., & Sasikumar, M. (2021). Power flow control based on bidirectional converter for hybrid power generation system using microcontroller. *Microprocessors and Microsystems*, *82*, 103950. DOI: 10.1016/j.micpro.2021.103950

Bhuiyan, M. R. A. (2022). Overcome the future environmental challenges through sustainable and renewable energy resources. *Micro & Nano Letters*, *17*(14), 402–416. DOI: 10.1049/mna2.12148

Bockris, J. O. M. (2013). The hydrogen economy : Its history. *International Journal of Hydrogen Energy*, *38*(6), 2579–2588. DOI: 10.1016/j.ijhydene.2012.12.026

Brandon, N. P., & Kurban, Z. (2017). Clean energy and the hydrogen economy. *Philosophical Transactions. Series A, Mathematical, Physical, and Engineering Sciences*, *375*(2098), 20160400. DOI: 10.1098/rsta.2016.0400 PMID: 28607181

Braun, K., & Kropp, C. (2023). Building a better world? Competing promises, visionsand imaginaries-in-the-making of the digitalization of architecture and construction. *Futures*, *154*, 103262. DOI: 10.1016/j.futures.2023.103262

Bridgeland, R., Chapman, A., Mclellan, B., Sofronis, P., & Fujii, Y. (2022). Challenges toward achieving a successful hydrogen economy in the US : Potential end-use and infrastructure analysis to the year 2100. *Cleaner Production Letters*, *3*, 100012. DOI: 10.1016/j.clpl.2022.100012

Britannica, Access date: 2025. https://www.britannica.com/science/hydrocarbon

Bromberg L. & Rabinovich A. (2001). Homogeneous charge compression ignition control by the use of plasmatron fuel converter technology. Massachusetts Institute of Technology Cambridge MA 02139.

Bromberg L., Cohn D.R., Heywood J. & Rabinovich A. (2002b). Onboard plasmatron generation of hydrogen rich gas for diesel engine exhaust aftertreatment and other applications. Massachusetts Institute of Technology Cambridge MA 02139.

Bromberg, L., Cohn, D. R., Heywood, J., & Rabinovich, A. (2002a). *Onboard plasmatron generation of hydrogen rich gas for diesel aftertreatment and other applications*. Plasma Science and Fusion Center Massachusetts Institute of Technology.

Bromberg, L., Cohn, D. R., Heywood, J., & Rabinovich, A. (2003b). *Hydrogen generation from plasmatron reformers and use for diesel exhaust aftertreatment*. Massachusetts Institute of Technology.

Bromberg, L., Cohn, D. R., & Rabinovich, A. (1999). *Aftertreatment of diesel vehicle emissions using compact plasmatron fuel converter-catalyst systems*. Plasma Science and Fusion Center Massachusetts Institute of Technology.

Bromberg, L., Cohn, D. R., Rabinovich, A., & Alexeev, N. (2000). *Hydrogen manufacturing using low current, non-thermal plasma boosted fuel converters*. Plasma Science and Fusion Center Massachusetts Institute of Technology.

Bromberg, L., Cohn, D. R., Rabinovich, A., Alexeev, N., Samokhin, A., Hadidi, K., Palaia, J., & Margarit-Bel, N. (2006). *Onboard plasmatron hydrogen production for improved vehicles*. Massachusetts Institute of Technology Plasma Science and Fusion Center.

Bromberg, L., Cohn, D. R., Rabinovich, A., & Heywood, J. (2001). Emissions reductions using hydrogen from plasmatron fuel converters. *International Journal of Hydrogen Energy*, *26*(10), 1115–1121. DOI: 10.1016/S0360-3199(01)00049-0

Bromberg, L., Crane, S., Rabinovich, A., Kong, Y., Cohn, D. R., Heywood, J., Alexeev, N., & Samokhin, A. (2003a). *Hydrogen generation from plasmatron reformers: a promising technology for NOx adsorber regeneration and other automotive applications. Diesel Engine Emission Reduction (DEER)*. Workshop.

Bryan, J., Meek, A., Dana, S., Sakir, M. S. I., & Wang, H. (2023). Modeling and design optimization of carbon-free hybrid energy systems with thermal and hydrogen storage. *International Journal of Hydrogen Energy*, *48*(99), 39097–39111. DOI: 10.1016/j.ijhydene.2023.03.135

Burchart, D., Gazda-Grzywacz, M., Grzywacz, P., Burmistrz, P., & Zarębska, K. (2023). Life Cycle Assessment of Hydrogen Production from Coal Gasification as an Alternative Transport Fuel. *Energies*, *16*(1), 383. Advance online publication. DOI: 10.3390/en16010383

Business Wire. (2023). Cement Industry in India 2022 - 2027. https://www.businesswire.com/news/home/20230123005601/en/India-Cement-Industry-Report-2022-Cement-Consumption-Reached-355.46-Million-Tons-in-FY-2022-and-is-Expected-to-Reach-450.78-Million-Tons-by-the-End-of-FY-2027---ResearchAndMarkets.com#:~:text=India%20is%20the%20second%2Dlargest,million%20tons%20in%20FY%202021. Accessed on: January 23, 2023.

Chauhan, C., Parida, V., & Dhir, A. (2022). Linking circular economy and digitalisation technologies: A systematic literature review of past achievements and future promises. *Technological Forecasting and Social Change*, *177*, 121508. DOI: 10.1016/j.techfore.2022.121508

Chel, A., & Kaushik, G. (2011). Renewable energy for sustainable agriculture. *Agronomy for Sustainable Development*, *31*(1), 91–118. DOI: 10.1051/agro/2010029

Cho, R. (2021). *Why We Need Green Hydrogen?*https://news.climate.columbia.edu/2021/01/07/need-green-hydrogen/

Chowdhury, S., Dey, P. K., Rodríguez-Espíndola, O., Parkes, G., Tuyet, N. T. A., Long, D. D., & Ha, T. P. (2022). Impact of organisational factors on the circular economy practices and sustainable performance of small and medium-sized enterprises in Vietnam. *Journal of Business Research*, *147*, 362–378. DOI: 10.1016/j.jbusres.2022.03.077

Cohn, D. R. (2003). *Cleaner, higher efficiency vehicles using plasmatrons*. Plasma Science and Fusion Center Massachusetts Institute of Technology.

Cohn, D. R. (2004). *Environmental protection and energy savings using plasma technology*. Plasma Science and Fusion Center Massachusetts Institute of Technology.

Cormos, C. C. (2023). Green hydrogen production from decarbonized biomass gasification: An integrated techno-economic and environmental analysis. *Energy*, *270*, 126926. DOI: 10.1016/j.energy.2023.126926

CSIDC. (2019). Cement Plant. *Chhattisgarh State Industrial Development Corporation.* https://csidc.in/core_sector/aluminum-plant-2/. Accessed on: October 10, 2023.

d'Hubert, X., Henry, F., & Poellaer, S. (2021). Benefits of Micro-grids/ Energy Storage for the Cement & Mineral Industries: Part II. The Green Hydrogen Option. In *2021 IEEE-IAS/PCA Cement Industry Conference (IAS/PCA)* (pp. 1-11). IEEE.

Dar, J., & Asif, M. (2023). Environmental feasibility of a gradual shift from fossil fuels to renewable energy in India: Evidence from multiple structural breaks cointegration. *Renewable Energy*, *202*, 589–601. DOI: 10.1016/j.renene.2022.10.131

de Oliveira, R. T., Ghobakhloo, M., & Figueira, S. (2023). Industry 4.0 Towards Social and Environmental Sustainability in Multinationals: Enabling Circular Economy, Organizational Social Practices and Corporate Purpose. *Journal of Cleaner Production*, *430*, 139712. DOI: 10.1016/j.jclepro.2023.139712

Dey, S., Sreenivasulu, A., Veerendra, G. T. N., Rao, K. V., & Babu, P. A. (2022). *Renewable energy present status and future potentials in India: An overview*. Innovation and Green Development. DOI: 10.1016/j.igd.2022.100006

Di Micco, S., Romano, F., Jannelli, E., Perna, A., & Minutillo, M. (2023). Techno-economic analysis of a multi-energy system for the co-production of green hydrogen, renewable electricity and heat. *International Journal of Hydrogen Energy*, *48*(81), 31457–31467. DOI: 10.1016/j.ijhydene.2023.04.269

Dillon, E. J., Hennessy, T., & Hynes, S. (2010). Assessing the sustainability of Irish agriculture. *International Journal of Agricultural Sustainability*, *8*(3), 131–147. DOI: 10.3763/ijas.2009.0044

Ding, J. W., Fu, Y. S., & Lisa Hsieh, I. Y. (2024). The cost of green: Analyzing the economic feasibility of hydrogen production from offshore wind power. Energy Conversion and Management: X, 24, 100770. DOI: 10.1016/j.ecmx.2024.100770

Eisapour, A. H., Jafarpur, K., & Farjah, E. (2021). Feasibility study of a smart hybrid renewable energy system to supply the electricity and heat demand of Eram Campus, Shiraz University; simulation, optimization, and sensitivity analysis. *Energy Conversion and Management*, *248*, 114779. DOI: 10.1016/j.enconman.2021.114779

Endo, N., Goshome, K., Tetsuhiko, M., Segawa, Y., Shimoda, E., & Nozu, T. (2021). Thermal management and power saving operations for improved energy efficiency within a renewable hydrogen energy system utilizing metal hydride hydrogen storage. *International Journal of Hydrogen Energy*, *46*(1), 262–271. DOI: 10.1016/j.ijhydene.2020.10.002

Energy, D. European Commission Directorate-General for Research Communication European Commission. High Level Group for Hydrogen, Fuel Cell Technologies, European Commission. Directorate-General for Energy, & European Commission. Directorate General for Research. (2003). Hydrogen Energy and Fuel Cells: A Vision of Our Future: [final Report of the High-Level Group] (Vol. 20719). European Commission, Directorate-General for Research. Unit B-1049 Brussels Fax (32-2) 29-58220.

Environmental Protection Agency. U. S. (2023). *Greenhouse Gases What are the trends in greenhouse gas emissions and concentrations and health and the environment?*https://www.epa.gov/report-environment/greenhouse-gases

Ewan, B. C. R., & Allen, R. W. K. (2005). A figure of merit assessment of the routes to hydrogen. *International Journal of Hydrogen Energy*, *30*(8), 809–819. DOI: 10.1016/j.ijhydene.2005.02.003

Fan, G., Liu, Z., Liu, X., Shi, Y., Wu, D., Guo, J., Zhang, S., Yang, X., & Zhang, Y. (2022). Two-layer collaborative optimization for a renewable energy system combining electricity storage, hydrogen storage, and heat storage. *Energy*, *259*, 125047. DOI: 10.1016/j.energy.2022.125047

Farrell, A. E., Keith, D. W., & Corbett, J. J. (2003). A strategy for introducing hydrogen into transportation. *Energy Policy*, *31*(13), 1357–1367. DOI: 10.1016/S0301-4215(02)00195-7

Furstenau, L. B., Sott, M. K., Kipper, L. M., Machado, E. L., Lopez-Robles, J. R., Dohan, M. S., Cobo, M. J., Zahid, A., Abbasi, Q. H., & Imran, M. A. (2020). Link between sustainability and industry 4.0: Trends, challenges and new perspectives. *IEEE Access : Practical Innovations, Open Solutions*, *8*, 140079–140096. DOI: 10.1109/ACCESS.2020.3012812

Garcia, M., & Oliva, S. (2023). Technical, economic, and CO2 emissions assessment of green hydrogen production from solar/wind energy: The case of Chile. *Energy*, *278*, 127981. DOI: 10.1016/j.energy.2023.127981

Gatell, I. S., & Avella, L. (2024). Impact of Industry 4.0 and circular economy on lean culture and leadership: Assessing digital green lean as a new concept. *European Research on Management and Business Economics*, *30*(1), 100232. DOI: 10.1016/j.iedeen.2023.100232

Ghribi, D., Khelifa, A., Diaf, S., & Belhamel, M. (2013). Study of hydrogen production system by using PV solar energy and PEM electrolyser in Algeria. *International Journal of Hydrogen Energy*, *38*(20), 8480–8490. DOI: 10.1016/j.ijhydene.2012.09.175

Gregory, O., Ng, D., & Long, G. (1972). The Electrochemistry of Cleaner Environments. In *Surface Electrochemistry* (pp. 927–976).

Guan, L., Li, W., Guo, C., & Huang, J. (2023). Environmental strategy for sustainable development: Role of digital transformation in China's natural resource exploitation. *Resources Policy*, *87*, 104304. DOI: 10.1016/j.resourpol.2023.104304

Hai, T., El-Shafay, A. S., Alizadeh, A. A., Dhahad, H. A., Chauhan, B. S., Almojil, S. F., Almohana, A. I., & Alali, A. F. (2023). Comparison analysis of hydrogen addition into both anode and afterburner of fuel cell incorporated with hybrid renewable energy driven SOFC: An application of techno-environmental horizon and multi-objective optimization. *International Journal of Hydrogen Energy*.

Han, F., Zeng, J., Lin, J., & Gao, C. (2023). Multi-stage distributionally robust optimization for hybrid energy storage in regional integrated energy system considering robustness and nonanticipativity. *Energy*, *277*, 127729. DOI: 10.1016/j.energy.2023.127729

Han, H., & Zhou, Z. (2024). The rebound effect of energy consumption and its determinants in China's agricultural production. *Energy, 290*, 129961. DOI: 10.1016/j.energy.2023.129961

Hart, D., Howes, J., Madden, B., & Boyd, E. (2016). *Hydrogen and Fuel Cells: Opportunities for Growth.*

Hassan, Q., Sameen, A. Z., Salman, H. M., & Jaszczur, M. (2023). Large-scale green hydrogen production via alkaline water electrolysis using solar and wind energy. *International Journal of Hydrogen Energy, 48*(88), 34299–34315. DOI: 10.1016/j.ijhydene.2023.05.126

Hassan, Q., Sameen, A. Z., Salman, H. M., Jaszczur, M., Al-Hitmi, M., & Alghoul, M. (2023). Energy futures and green hydrogen production: Is Saudi Arabia trend? *Results in Engineering, 18*, 101165. DOI: 10.1016/j.rineng.2023.101165

HassanzadehFard, H., Tooryan, F., & Dargahi, V. (2023). Standalone hybrid system energy management optimization for remote village considering methane production from livestock manure. *International Journal of Hydrogen Energy, 48*(29), 10778–10796. DOI: 10.1016/j.ijhydene.2022.12.085

He, J., Wu, Y., Wu, M., Xu, M., & Liu, F. (2022). Two-stage configuration optimization of a novel standalone renewable integrated energy system coupled with hydrogen refueling. *Energy Conversion and Management, 251*, 114953. DOI: 10.1016/j.enconman.2021.114953

He, J., Wu, Y., Yong, X., Tan, Q., & Liu, F. (2022). Bi-level optimization of a near-zero-emission integrated energy system considering electricity-hydrogen-gas nexus: A two-stage framework aiming at economic and environmental benefits. *Energy Conversion and Management, 274*, 116434. DOI: 10.1016/j.enconman.2022.116434

Henry, A., McStay, D., Rooney, D., Robertson, P., & Foley, A. (2023). Techno-economic analysis to identify the optimal conditions for green hydrogen production. *Energy Conversion and Management, 291*, 117230. DOI: 10.1016/j.enconman.2023.117230

Herdem, M. S., Mazzeo, D., Matera, N., Baglivo, C., Khan, N., Congedo, P. M., & DeGiorgi, M. G. (2023). A brief overview of solar and wind-based green hydrogen production systems: Trends and standardization. *International Journal of Hydrogen Energy*.

Hettiarachchi, B. D., Brandenburg, M., & Seuring, S. (2022). Connecting additive manufacturing to circular economy implementation strategies: Links, contingencies and causal loops. *International Journal of Production Economics, 246*, 108414. DOI: 10.1016/j.ijpe.2022.108414

He, Y., Guo, S., Dong, P., Lv, D., & Zhou, J. (2023). Feasibility analysis of decarbonizing coal-fired power plants with 100% renewable energy and flexible green hydrogen production. *Energy Conversion and Management, 290*, 117232. DOI: 10.1016/j.enconman.2023.117232

Huangfu, Y., Tian, C., Zhuo, S., Xu, L., Li, P., Quan, S., Zhang, Y., & Ma, R. (2023). An optimal energy management strategy with subsection bi-objective optimization dynamic programming for photovoltaic/battery/hydrogen hybrid energy system. *International Journal of Hydrogen Energy, 48*(8), 3154–3170. DOI: 10.1016/j.ijhydene.2022.10.133

Huang, J., Balcombe, P., & Feng, Z. (2023). Technical and economic analysis of different colours of producing hydrogen in China. *Fuel, 337*, 127227. DOI: 10.1016/j.fuel.2022.127227

International Energy Agency. (2023). *World Energy Outlook 2023 | Enhanced Reader*. https://www.iea.org/news/the-energy-world-is-set-to-change-significantly-by-2030-based-on-today-s-policy-settings-alone

IREA. (2022). Hydrogen from Renewable Power. *International Renewable Energy Agency*. https://www.irena.org/Energy-Transition/Technology/Hydrogen. Accessed on: 16 October 2023.

Ishaq, H., & Dincer, I. (2021). Comparative assessment of renewable energy-based hydrogen production methods. *Renewable & Sustainable Energy Reviews, 135*, 110192. DOI: 10.1016/j.rser.2020.110192

Izadi, A., Shahafve, M., & Ahmadi, P. (2022). Neural network genetic algorithm optimization of a transient hybrid renewable energy system with solar/wind and hydrogen storage system for zero energy buildings at various climate conditions. *Energy Conversion and Management*, *260*, 115593. DOI: 10.1016/j.enconman.2022.115593

Izadi, A., Shahafve, M., Ahmadi, P., & Hanafizadeh, P. (2023). Design, and optimization of COVID-19 hospital wards to produce Oxygen and electricity through solar PV panels with hydrogen storage systems by neural network-genetic algorithm. *Energy*, *263*, 125578. DOI: 10.1016/j.energy.2022.125578 PMID: 36189102

James, A. T., Kumar, G., James, J., & Asjad, M. (2023). Development of a micro-level circular economy performance measurement framework for automobile maintenance garages. *Journal of Cleaner Production*, *417*, 138025. DOI: 10.1016/j.jclepro.2023.138025

Jat, H. S., Datta, A., Choudhary, M., Sharma, P. C., & Jat, M. L. (2021). Conservation Agriculture: Factors and drivers of adoption and scalable innovative practices in Indo-Gangetic plains of India–a review. *International Journal of Agricultural Sustainability*, *19*(1), 40–55. DOI: 10.1080/14735903.2020.1817655

Jawerth, N. (2022). *What is the Clean Energy Transition and How Does Nuclear Power Fit In?*https://www.iaea.org/bulletin/what-is-the-clean-energy-transition-and-how-does-nuclear-power-fit-in

Jia, K., Liu, C., Li, S., & Jiang, D. (2023). Modeling and optimization of a hybrid renewable energy system integrated with gas turbine and energy storage. *Energy Conversion and Management*, *279*, 116763. DOI: 10.1016/j.enconman.2023.116763

Jindal, A., Shrimali, G., & Tiwary, N. (2024). At scale adoption of Green Hydrogen in Indian Industry: Costs, subsidies and policies. *Energy for Sustainable Development*, *83*, 101549. DOI: 10.1016/j.esd.2024.101549

Ji, W., Zhang, K., Zhan, K., Wang, P., Wang, X., & Yan, Y. (2022). Solar-Powered Environmentally Friendly Hydrogen Production: Advanced Technologies for Sunlight- Electricity-Hydrogen Nexus. *Chinese Journal of Structural Chemistry*, *41*(5), 2205015–2205029.

Juangsa, F. B., Cezeliano, A. S., Darmanto, P. S., & Aziz, M. (2022). Thermodynamic analysis of hydrogen utilization as alternative fuel in cement production. *South African Journal of Chemical Engineering*, *42*, 23–31. DOI: 10.1016/j.sajce.2022.07.003

Jurado, F., Mezrhab, A., Moussaoui, M. A., & Vera, D. (2023). Cost and size optimization of hybrid solar and hydrogen subsystem using HomerPro software. *International Journal of Hydrogen Energy*.

Karayel, G. K., Javani, N., & Dincer, I. (2022). Green hydrogen production potential for Turkey with solar energy. *International Journal of Hydrogen Energy*, *47*(45), 19354–19364. DOI: 10.1016/j.ijhydene.2021.10.240

Khan, D. A., Choudhary, A. K., & Sharma, D. (2024). *Hydrogen Energy*. CRC Press. DOI: 10.1201/9781003537816

Khan, T., Yu, M., & Waseem, M. (2022). Review on recent optimization strategies for hybrid renewable energy system with hydrogen technologies: State of the art, trends and future directions. *International Journal of Hydrogen Energy*, *47*(60), 25155–25201. DOI: 10.1016/j.ijhydene.2022.05.263

Kilic, M., & Altun, A. F. (2022). Dynamic modelling and multi-objective optimization of off-grid hybrid energy systems by using battery or hydrogen storage for different climates. *International Journal of Hydrogen Energy*.

Knickel, K., Ashkenazy, A., Chebach, T. C., & Parrot, N. (2017). Agricultural modernization and sustainable agriculture: Contradictions and complementarities. *International Journal of Agricultural Sustainability*, *15*(5), 575–592. DOI: 10.1080/14735903.2017.1373464

Kochanek, E. (2022). The Role of Hydrogen in the Visegrad Group Approach to Energy Transition. *Energies*, *15*(19), 7235. https://doi.org/. DOI: 10.3390/en15197235

Koohafkan, P., Altieri, M. A., & Gimenez, E. H. (2012). Green agriculture: Foundations for biodiverse, resilient and productive agricultural systems. *International Journal of Agricultural Sustainability, 10*(1), 61–75. DOI: 10.1080/14735903.2011.610206

Kumar, S. S., & Lim, H. (2022). An overview of water electrolysis technologies for green hydrogen production. *Energy Reports*, *8*, 13793–1381. DOI: 10.1016/j.egyr.2022.10.127

Kumar, S., & Kumar, K. R. (2022). Techno economic feasibility study on hydrogen production using concentrating solar thermal technology in India. *International Journal of Hydrogen Energy*, *47*(89), 37708–37723. DOI: 10.1016/j.ijhydene.2022.08.285

Kurniawan, T. A., Maiurova, A., Kustikova, M., Bykovskaia, E., Othman, M. H. D., & Goh, H. H. (2022). Accelerating sustainability transition in St. Petersburg (Russia) through digitalization-based circular economy in waste recycling industry: A strategy to promote carbon neutrality in era of Industry 4.0. *Journal of Cleaner Production*, *363*, 132452. DOI: 10.1016/j.jclepro.2022.132452

Kurniawan, T. A., Othman, M. H. D., Hwang, G. H., & Gikas, P. (2022). Unlocking digital technologies for waste recycling in Industry 4.0 era: A transformation towards a digitalization-based circular economy in Indonesia. *Journal of Cleaner Production*, *357*, 131911. DOI: 10.1016/j.jclepro.2022.131911

Kurniawan, T. A., Othman, M. H. D., Liang, X., Goh, H. H., Gikas, P., Kusworo, T. D., Anouzla, A., & Chew, K. W. (2023). Decarbonization in waste recycling industry using digitalization to promote net-zero emissions and its implications on sustainability. *Journal of Environmental Management*, *338*, 117765. DOI: 10.1016/j.jenvman.2023.117765 PMID: 36965421

Lamb, W. F., Wiedmann, T., Pongratz, J., Andrew, R., Crippa, M., Olivier, J. G. J., Wiedenhofer, D., Mattioli, G., Al Khourdajie, A., & House, J. (2021). A review of trends and drivers of greenhouse gas emissions by sector from 1990 to 2018. *Environmental Research Letters*, *16*(7), 073005. https://doi.org/https://doi.org/10.1088/1748-9326/abee4e. DOI: 10.1088/1748-9326/abee4e

Lee, S., Na, U. J., & Jo, H. (2023). Techno-economic assessment of green hydrogen production via two-step thermochemical water splitting using microwave. *International Journal of Hydrogen Energy*, *48*(29), 10706–10723. DOI: 10.1016/j.ijhydene.2022.12.119

Li, J., Liu, P., & Li, Z. (2022). Optimal design and techno-economic analysis of a hybrid renewable energy system for off-grid power supply and hydrogen production: A case study of West China. *Chemical Engineering Research & Design*, *177*, 604–614. DOI: 10.1016/j.cherd.2021.11.014

Liu, J., Zhou, Y., Yang, H., & Wu, H. (2022). Net-zero energy management and optimization of commercial building sectors with hybrid renewable energy systems integrated with energy storage of pumped hydro and hydrogen taxis. *Applied Energy*, *321*, 119312. DOI: 10.1016/j.apenergy.2022.119312

Liu, Y., Yu, Y., Huang, Y., & Guan, W. (2024). Utilizing the resources efficiency: Evidence from the impacts of media industry and digitalization. *Resources Policy*, *88*, 104346. DOI: 10.1016/j.resourpol.2023.104346

Live Mint. (2023). India rolls out green hydrogen production standards. *Live Mint*. https://www.livemint.com/industry/energy/india-rolls-out-green-hydrogen-production-standards-11692426710928.html. Accessed on: 19 Aug 2023

Li, X., & Mulder, M. (2021). Value of power-to-gas as a flexibility option in integrated electricity and hydrogen markets. *Applied Energy*, *304*, 117863. DOI: 10.1016/j.apenergy.2021.117863

Lu, H., Xi, D., & Cheng, Y. F. (2025). Hydrogen production in integration with CCUS: A realistic strategy towards net zero. *Energy*, *315*, 134398. DOI: 10.1016/j.energy.2025.134398

Luo, L., Cristofari, C., & Levrey, S. (2023). Cogeneration: Another way to increase energy efficiency of hybrid renewable energy hydrogen chain–A review of systems operating in cogeneration and of the energy efficiency assessment through exergy analysis. *Journal of Energy Storage*, *66*, 107433. DOI: 10.1016/j.est.2023.107433

Lu, W. (2017). Greenhouse Gas Emissions, Energy Consumption and Economic Growth : A Panel Cointegration Analysis for 16 Asian Countries. *International Journal of Environmental Research and Public Health*, *14*(11), 1436. DOI: 10.3390/ijerph14111436 PMID: 29165399

Maciel, L. B. B., Viola, L., de Queiróz Lamas, W., & Silveira, J. L. (2023). Environmental studies of green hydrogen production by electrolytic process: A comparison of the use of electricity from solar PV, wind energy, and hydroelectric plants. *International Journal of Hydrogen Energy*, *48*(93), 36584–36604. DOI: 10.1016/j.ijhydene.2023.05.334

Maestre, V. M., Ortiz, A., & Ortiz, I. (2022). Transition to a low-carbon building stock. Techno-economic and spatial optimization of renewables-hydrogen strategies in Spain. *Journal of Energy Storage*, *56*, 105889. DOI: 10.1016/j.est.2022.105889

Matt, D. T., Pedrini, G., Bonfanti, A., & Orzes, G. (2023). Industrial digitalization. A systematic literature review and research agenda. *European Management Journal*, *41*(1), 47–78. DOI: 10.1016/j.emj.2022.01.001

Mazzeo, D., Herdem, M. S., Matera, N., & Wen, J. Z. (2022). Green hydrogen production: Analysis for different single or combined large-scale photovoltaic and wind renewable systems. *Renewable Energy*, *200*, 360–378. DOI: 10.1016/j.renene.2022.09.057

Mcdowall, W. (2012). Technology roadmaps for transition management : The case of hydrogen energy. *Technological Forecasting and Social Change*, *79*(3), 530–542. DOI: 10.1016/j.techfore.2011.10.002

Mcdowall, W. (2014). Exploring possible transition pathways for hydrogen energy : A hybrid approach using socio-technical scenarios and energy system modelling. *Futures*, *63*, 1–14. DOI: 10.1016/j.futures.2014.07.004

Mehrenjani, J. R., Gharehghani, A., Nasrabadi, A. M., & Moghimi, M. (2022). Design, modeling and optimization of a renewable-based system for power generation and hydrogen production. *International Journal of Hydrogen Energy*, *47*(31), 14225–14242. DOI: 10.1016/j.ijhydene.2022.02.148

Ministry of Environment. Forest and Climate Change (2023). Net zero emissions target. https://pib.gov.in/PressReleaseIframePage.aspx?PRID=1945472#:~:text=India%2C%20at%20the%2026th%20session,achieve%20net%20zero%20by%202070

Mint, (August 14, 2023). *Indian agriculture must balance profits with sustainable practices.*https://www.livemint.com/opinion/first-person/indian-agriculture-must-balance-profits-with-sustainable-practices-11691889108165.html (Accessed on: December 18, 2023).

Modgil, S., Gupta, S., Sivarajah, U., & Bhushan, B. (2021). Big data-enabled large-scale group decision making for circular economy: An emerging market context. *Technological Forecasting and Social Change*, *166*, 120607. DOI: 10.1016/j.techfore.2021.120607

Mohamed, B., Ali, B., Ahmed, B., Ahmed, B., Salah, L., & Rachid, D. (2016). Study of hydrogen production by solar energy as tool of storing and utilization renewable energy for the desert areas. *International Journal of Hydrogen Energy*, *41*(45), 20788–20806. DOI: 10.1016/j.ijhydene.2016.07.034

Mohd Amin, N. A. A., & Mohd Zaid, H. F. (2024). A review of hydrogen production using $TIO_2$-based photocatalyst in tandem solar cell. *International Journal of Hydrogen Energy*, *77*, 166–183. DOI: 10.1016/j.ijhydene.2024.05.407

Müller, L. A., Leonard, A., Trotter, P. A., & Hirmer, S. (2023). Green hydrogen production and use in low-and middle-income countries: A least-cost geospatial modelling approach applied to Kenya. *Applied Energy*, *343*, 121219. DOI: 10.1016/j.apenergy.2023.121219

Munir, M. T., Li, B., Naqvi, M., & Nizami, A. S. (2023). Green loops and clean skies: Optimizing municipal solid waste management using data science for a circular economy. *Environmental Research*, •••, 117786. PMID: 38036215

Murray, L. J., Dincă, M., & Long, J. R. (2009). Hydrogen storage in metal–organic frameworks. *Chemical Society Reviews*, *38*(5), 1294–1314. DOI: 10.1039/b802256a PMID: 19384439

Nasser, M., & Hassan, H. (2023). Egyptian green hydrogen Atlas based on available wind/solar energies: Power, hydrogen production, cost, and CO2 mitigation maps. *International Journal of Hydrogen Energy*.

Nikkei. (2013). *Overview of World Hydrogen Infrastructure Projects*.

Nithyanandan, S., & Sindhu Vaardini, U. (2023). A Review on Implementing Lean Construction Technique in Construction Industry. *International Journal of Advanced Research in Science*. *Tongxin Jishu*, •••, 791–796.

Noussan, M., Raimondi, P. P., Scita, R., & Hafner, M. (2020). The Role of Green and Blue Hydrogen in the Energy Transition - A Technological and Geopolitical Perspective. *Sustainability (Basel)*, *13*(1), 298. DOI: 10.3390/su13010298

Osman, A. I., Mehta, N., Elgarahy, A. M., Hefny, M., Al-Hinai, A., Al-Muhtaseb, A. A. H., & Rooney, D. W. (2022). Hydrogen production, storage, utilisation and environmental impacts: A review. *Environmental Chemistry Letters*, *20*(1), 1–36. DOI: 10.1007/s10311-021-01322-8

Özsungur, F., Mortaza, C. S., & Hülya, K. B. (2025). *Geopolitical Landscapes of Renewable Energy and Urban Growth*. IGI Global. DOI: 10.4018/979-8-3693-8814-3

Pal, P., & Mukherjee, V. (2021). Off-grid solar photovoltaic/hydrogen fuel cell system for renewable energy generation: An investigation based on techno-economic feasibility assessment for the application of end-user load demand in North-East India. *Renewable & Sustainable Energy Reviews*, *149*, 111421. DOI: 10.1016/j.rser.2021.111421

Pei, Z., Yu, T., Yi, W., & Li, Y. (2021). Twenty-year retrospection on green manufacturing: A bibliometric perspective. *IET Collaborative Intelligent Manufacturing*, *3*(4), 303–323. DOI: 10.1049/cim2.12038

Pocha, C. K. R., Chia, W. Y., Kurniawan, T. A., Khoo, K. S., & Chew, K. W. (2023). Thermochemical conversion of different biomass feedstocks into hydrogen for power plant electricity generation. *Fuel*, *340*, 127472. DOI: 10.1016/j.fuel.2023.127472

Premadasa, P. N. D., Silva, C. M. M. R. S., Chandima, D. P., & Karunadasa, J. P. (2023). A multi-objective optimization model for sizing an off-grid hybrid energy microgrid with optimal dispatching of a diesel generator. *Journal of Energy Storage*, *68*, 107621. DOI: 10.1016/j.est.2023.107621

Pretty, J., Smith, G., Goulding, K. W. T., Groves, S. J., Henderson, I., Hine, R. E., King, V., Van Oostrum, J., Pendlington, D. J., Vis, J. K., & Walter, C. (2008). Multi-year assessment of Unilever's progress towards agricultural sustainability II: Outcomes for peas (UK), spinach (Germany, Italy), tomatoes (Australia, Brazil, Greece, USA), tea (Kenya, Tanzania, India) and oil palm (Ghana). *International Journal of Agricultural Sustainability*, *6*(1), 63–88. DOI: 10.3763/ijas.2007.0323

Qiu, R., Zhang, H., Wang, G., Liang, Y., & Yan, J. (2023). Green hydrogen-based energy storage service via power-to-gas technologies integrated with multi-energy microgrid. *Applied Energy*, *350*, 121716. DOI: 10.1016/j.apenergy.2023.121716

Rajkumar, P., Anand, B. P., Varuvel, E. G., Saravanan, C. G., Josephin, J. S. F., Chinnathambi, A., & Pugazhendhi, A. (2025). Hydrogen-enriched lean-burn strategy for gasoline direct injection engine utilizing non-thermal plasma fuel reformer. *International Journal of Hydrogen Energy*, *107*, 512–525. DOI: 10.1016/j.ijhydene.2024.11.070

Renewable, I., & Energy Agency. (2020). *Green Hydrogen: A guide to policy making*.

Ren, X., Dong, L., Xu, D., & Hu, B. (2020). Challenges towards hydrogen economy in China. *International Journal of Hydrogen Energy*, *45*(59), 34326–34345. DOI: 10.1016/j.ijhydene.2020.01.163

Rezaei, M., Akimov, A., & Gray, E. M. A. (2024). Cost-competitiveness of green hydrogen and its sensitivity to major financial and technical variables. *International Journal of Hydrogen Energy*, *90*, 1365–1377. DOI: 10.1016/j.ijhydene.2024.10.068

Ritala, P., Baiyere, A., Hughes, M., & Kraus, S. (2021). Digital strategy implementation: The role of individual entrepreneurial orientation and relational capital. *Technological Forecasting and Social Change*, *171*, 120961. DOI: 10.1016/j.techfore.2021.120961

Rosen, M. A., & Koohi-fayegh, S. (2016). The prospects for hydrogen as an energy carrier : An overview of hydrogen energy and hydrogen energy systems. *Energy, Ecology & Environment*, *1*(1), 10–29. DOI: 10.1007/s40974-016-0005-z

Ruhnau, O. (2022). How flexible electricity demand stabilizes wind and solar market values: The case of hydrogen electrolyzers. *Applied Energy*, *307*, 118194. DOI: 10.1016/j.apenergy.2021.118194

S&P Global Commodities Insights. (2023). Global hydrogen demand to more than triple by 2050, driven by power generation, aviation and industry: analyst. https://www.hydrogeninsight.com/production/global-hydrogen-demand-to-more-than-triple-by-2050-driven-by-power-generation-aviation-and-industry-analyst/2-1-1394924

Saedi, I., Mhanna, S., & Mancarella, P. (2021). Integrated electricity and gas system modelling with hydrogen injections and gas composition tracking. *Applied Energy*, *303*, 117598. DOI: 10.1016/j.apenergy.2021.117598

Sahoo, N., Kumar, A., & Samsher, . (2023). Design of solar cement plant for supplying thermal energy in cement production. *Journal of Cleaner Production*, *426*, 139151. DOI: 10.1016/j.jclepro.2023.139151

Sandaka, B. P., & Kumar, J. (2023). Alternative vehicular fuels for environmental decarbonization: A critical review of challenges in using electricity, hydrogen, and biofuels as a sustainable vehicular fuel. *Chemical Engineering Journal Advances*, *14*, 100442. DOI: 10.1016/j.ceja.2022.100442

Satapathy, S., & Chauhan, H. (2024). Sustainable Supply Chain Management in the Mining Industry. DOI: 10.1201/9781003488347

Satapathy, S. (2024). Sustainable Supply Chain Management Challenges in Petroleum Transportation. *Advances in Logistics, Operations, and Management Science*, ●●●, 182–205. DOI: 10.4018/979-8-3693-0669-7.ch009

Satyapal, S. (2017). *Hydrogen: A Clean, Flexible Energy Carrier*. https://www.energy.gov/eere/articles/hydrogenclean-flexible-energy-carrier

Searle, S., & Zhou, Y. (2021). *Don't let the industry greenwash green hydrogen.*

Shahverdian, M. H., Sedayevatan, S., Hosseini, M., Sohani, A., Javadijam, R., & Sayyaadi, H. (2023). Multi-objective technoeconomic optimization of an off-grid solar-ground-source driven cycle with hydrogen storage for power and fresh water production. *International Journal of Hydrogen Energy*, *48*(52), 19772–19791. DOI: 10.1016/j.ijhydene.2023.02.062

Sharma, H. B., Vanapalli, K. R., Samal, B., Cheela, V. S., Dubey, B. K., & Bhattacharya, J. (2021). Circular economy approach in solid waste management system to achieve UN-SDGs: Solutions for post-COVID recovery. *The Science of the Total Environment*, *800*, 149605. DOI: 10.1016/j.scitotenv.2021.149605 PMID: 34426367

Sharma, R., Sharda, H., Dutta, A., Dahiya, A., Chaudhary, R., Singh, A., Rathi, K., Kumar, S., Sharma, A., Maken, S., & Nehra, S. P. (2023). Optimizing green hydrogen production: Leveraging load profile simulation and renewable energy integration. *International Journal of Hydrogen Energy*, *48*(96), 38015–38026. DOI: 10.1016/j.ijhydene.2023.03.179

Shin, H., Jang, D., Lee, S., Cho, H. S., Kim, K. H., & Kang, S. (2023). Techno-economic evaluation of green hydrogen production with low-temperature water electrolysis technologies directly coupled with renewable power sources. *Energy Conversion and Management*, *286*, 117083. DOI: 10.1016/j.enconman.2023.117083

Simona, R. M., Saheb, D., Omar, M., Candin, M. T., & Mustapha, K. (2024). *Advancements in Renewable Energy and Green Hydrogen*. IGI Global.

Singh, R., Altaee, A., & Gautam, S. (2020). Nanomaterials in the advancement of hydrogen energy storage. *Heliyon*, *6*(7), e04487. DOI: 10.1016/j.heliyon.2020.e04487 PMID: 32743097

Sisternes, F. de, & Jackson, C. (2020). *Green Hydrogen in develoing contries.*

Soni, N., & Reddy, K. S. (2024). Techno-economic assessment of concentrated solar power technologies integrated with thermal energy storage system for green hydrogen production. *International Journal of Hydrogen Energy*, *72*, 1184–1203. DOI: 10.1016/j.ijhydene.2024.05.409

Sousa, V., Bogas, J. A., Real, S., Meireles, I., & Carriço, A. (2023). Recycled cement production energy consumption optimization. *Sustainable Chemistry and Pharmacy*, *32*, 101010. DOI: 10.1016/j.scp.2023.101010

Sun, L., Li, X., & Wang, Y. (2024). Digital trade growth and mineral resources In developing countries: Implications for green recovery. *Resources Policy*. 88, pp. 104338. S&P Global, (August 3, 2023). *Future Farming: Agriculture's Role in a More Sustainable India.* https://www.spglobal.com/en/research-insights/featured/special-editorial/look-forward/future-farming-agriculture-s-role-in-a-more-sustainable-india (Accessed on: December 18, 2023).

Tarhan, C., & Çil, M. A. (2021). A study on hydrogen, the clean energy of the future: Hydrogen storage methods. *Journal of Energy Storage*, *40*, 102676. DOI: 10.1016/j.est.2021.102676

Thapa, B. S., Neupane, B., Yang, H., & Lee, Y.-H. (2021). Green hydrogen potentials from surplus hydro energy in Nepal. *International Journal of Hydrogen Energy*, *46*(43), 22256–22267. DOI: 10.1016/j.ijhydene.2021.04.096

The Energy and Resources Institute. (2023). India transforming to a net-zero emissions energy system - A call to action to 2030. chrome-extension://efaidnbmnnnibpcajpcglclefindmkaj/https://teriin.org/sites/default/files/files/India-Scenarios-Sketch.pdf

The Indian Express. (2023). Green Hydrogen push will need to counter challenges. https://indianexpress.com/article/opinion/editorials/green-hydrogen-push-will-need-to-counter-challenges-8957673/

The News India Express. (August 18, 2023). *India's agriculture sustainability at risk.* https://www.newindianexpress.com/xplore/2023/aug/18/indiasagriculture-sustainability-at-risk-2606814.html (Accessed on: December 18, 2023).

The Sunday Guardian. (October 14, 2023). *The Green Revolution in India: Harnessing Electric Tractors for Sustainable Agriculture.* https://sundayguardianlive.com/business/the-green-revolution-in-india-harnessing-electric-tractors-for-sustainable-agriculture (Accessed on: December 18, 2023).

The Times of India. (May 4, 2023). *Embracing sustainable agriculture: A virtuous cycle of prosperity.* https://timesofindia.indiatimes.com/blogs/voices/embracing-sustainable-agriculture-a-virtuous-cycle-of-prosperity/ (Accessed on: December 18, 2023).

The Wire. (July 21, 2023). *How Natural Farming Can Revive India's Farmlands and Ensure Sustainable Agriculture.*https://thewire.in/agriculture/how-natural-farming-can-revive-indias-farmlands-and-ensure-sustainable-agriculture (Accessed on: December 18, 2023).

Tleubergenova, A., Abuov, Y., Danenova, S., Khoyashov, N., Togay, A., & Lee, W. (2023). Resource assessment for green hydrogen production in Kazakhstan. *International Journal of Hydrogen Energy*, *48*(43), 16232–16245. DOI: 10.1016/j.ijhydene.2023.01.113

Torky, M., Dahy, G., & Hassanein, A. E. (2023). GH2_MobileNet: Deep learning approach for predicting green hydrogen production from organic waste mixtures. *Applied Soft Computing*, *138*, 110215. DOI: 10.1016/j.asoc.2023.110215

Truant, E., Giordino, D., Borlatto, E., & Bhatia, M. (2024). Drivers and barriers of smart technologies for circular economy: Leveraging smart circular economy implementation to nurture companies' performance. *Technological Forecasting and Social Change, 198*, 122954. DOI: 10.1016/j.techfore.2023.122954

Tuluhong, A., Chang, Q., Xie, L., Xu, Z., & Song, T. (2024). Current Status of Green Hydrogen Production Technology: A Review. In Sustainability (Switzerland) 16(20). DOI: 10.3390/su16209070

UN DESA. (2023). UN Department of Economic and Social Affairs (DESA) Policy Brief No. 153: India overtakes China as the world's most populous country, https://www.un.org/development/desa/dpad/publication/un-desa-policy-brief-no-153-india-overtakes-china-as-the-worlds-most-populous-country/

Upadhyay, A., Balodi, K. C., Naz, F., Di Nardo, M., & Jraisat, L. (2023). Implementing industry 4.0 in the manufacturing sector: Circular economy as a societal solution. *Computers & Industrial Engineering*, *177*, 109072. DOI: 10.1016/j.cie.2023.109072

Urunkar, R. U., & Patil, S. D. (2021b). Numerical study on performance evaluation of multi-tubular sodium alanate hydride reactor by enhancing heat and mass transfer characteristics using nanofluids. *Proceedings of the 26th National and 4th International ISHMT-ASTFE Heat and Mass Transfer ConferenceDecember 17-20, 2021, IIT Madras, Chennai-600036, Tamil Nadu, India*, 2613–2618. DOI: 10.1615/IHMTC-2021.3950

Urunkar, R. U., & Patil, S. D. (2021a). Enhancement of heat and mass transfer characteristics of metal hydride reactor for hydrogen storage using various nanofluids. *International Journal of Hydrogen Energy*, *46*(37), 19486–19497. DOI: 10.1016/j.ijhydene.2021.03.090

Urunkar, R. U., & Patil, S. D. (2023a). Hydrogen Storage Technologies and Related Heat and Mass Transfer Studies. In *Hydrogen Fuel Cell Technology for Mobile Applications* (pp. 182–206). IGI Global., DOI: 10.4018/978-1-6684-6721-3.ch008

Urunkar, R. U., & Patil, S. D. (2023b). Performance analysis of sodium alanate hydride reactor with different nanofluids. *International Journal of Hydrogen Energy*, *48*(96), 37944–37953. DOI: 10.1016/j.ijhydene.2023.02.105

Van Der Zwaan, B., Lamboo, S., & Dalla Longa, F. (2021). Timmermans' dream: An electricity and hydrogen partnership between Europe and North Africa. *Energy Policy*, *159*, 112613. DOI: 10.1016/j.enpol.2021.112613

Vatican News. (December 04, 2023). *COP28: Sustainable agriculture is antidote to climate change.*https://www.vaticannews.va/en/world/news/2023-12/cop28-sustainable-agriculture-antidote-to-climate-change.html (Accessed on: December 18, 2023).

Vives, A. M. V., Wang, R., Roy, S., & Smallbone, A. (2023). Techno-economic analysis of large-scale green hydrogen production and storage. *Applied Energy*, *346*, 121333. DOI: 10.1016/j.apenergy.2023.121333

Walter, V., Göransson, L., Taljegard, M., Öberg, S., & Odenberger, M. (2023). Low-cost hydrogen in the future European electricity system–Enabled by flexibility in time and space. *Applied Energy*, *330*, 120315. DOI: 10.1016/j.apenergy.2022.120315

Wang, H., Zhao, X., Zhang, K., & Wang, W. (2022). Economic assessment of a renewable energy-electricity-hydrogen system considering environmental benefits. *Sustainable Production and Consumption*, *33*, 903–920. DOI: 10.1016/j.spc.2022.08.010

Wang, R., & Zhang, R. (2023). Techno-economic analysis and optimization of hybrid energy systems based on hydrogen storage for sustainable energy utilization by a biological-inspired optimization algorithm. *Journal of Energy Storage*, *66*, 107469. DOI: 10.1016/j.est.2023.107469

Wang, Z., Huang, Y., Ankrah, V., & Dai, J. (2023). Greening the knowledge-based economies: Harnessing natural resources and innovation in information and communication technologies for green growth. *Resources Policy*, *86*, 104181. DOI: 10.1016/j.resourpol.2023.104181

World Health Organization. (2018). COP24 special report: health and climate change. Access date: 2020. https://www.energy.gov/eere/articles/hydrogen-clean-flexible-energy-carrier

Wu, S., Lin, J., Li, J., & Song, Y. (2022). Electricity allocation strategy for on-site hydrogen refueling stations in the forward and spot markets. *IET Renewable Power Generation*, *16*(12), 2532–2546. DOI: 10.1049/rpg2.12406

Xu, Z., Shen, Y., Chen, L., & Wang, C. (2025). High $SO_2$ tolerant and long term reliability of MnSO4-coated Ni-based catalysts for $CO_2$ methanation in coal fired flue gas. *Chemical Engineering Journal*, *507*, 160820. DOI: 10.1016/j.cej.2025.160820

Yang, G., Zhang, H., Wang, W., Liu, B., Lyu, C., & Yang, D. (2023). Capacity optimization and economic analysis of PV–hydrogen hybrid systems with physical solar power curve modeling. *Energy Conversion and Management*, *288*, 117128. DOI: 10.1016/j.enconman.2023.117128

Yang, J., Sudik, A., Wolverton, C., & Siegel, D. J. (2010). High capacity hydrogen storage materials: Attributes for automotive applications and techniques for materials discovery. *Chemical Society Reviews*, *39*(2), 656–675. DOI: 10.1039/B802882F PMID: 20111786

Yan, R., Wang, J., Huo, S., Qin, Y., Zhang, J., Tang, S., Wang, Y., Liu, Y., & Zhou, L. (2023). Flexibility improvement and stochastic multi-scenario hybrid optimization for an integrated energy system with high-proportion renewable energy. *Energy*, *263*, 125779. DOI: 10.1016/j.energy.2022.125779

Yue, M., Lambert, H., Pahon, E., Roche, R., Jemei, S., & Hissel, D. (2021). Hydrogen energy systems: A critical review of technologies, applications, trends and challenges. *Renewable & Sustainable Energy Reviews*, *146*, 111180. DOI: 10.1016/j.rser.2021.111180

Zhang, H., & Xie, Y. (2024). Assessing natural resources, rebounding trends, digital economic structure and green recovery dynamics in China. *Resources Policy*, *88*, 104482. DOI: 10.1016/j.resourpol.2023.104482

Zhang, K., Zhou, B., Chung, C. Y., Bu, S., Wang, Q., & Voropai, N. (2022). A coordinated multi-energy trading framework for strategic hydrogen provider in electricity and hydrogen markets. *IEEE Transactions on Smart Grid, 14*(2), 1403–1417. DOI: 10.1109/TSG.2022.3154611

Zhang, K., Zhou, B., Or, S. W., Li, C., Chung, C. Y., & Voropai, N. (2021). Optimal coordinated control of multi-renewable-to-hydrogen production system for hydrogen fueling stations. *IEEE Transactions on Industry Applications*, *58*(2), 2728–2739. DOI: 10.1109/TIA.2021.3093841

Zhang, L., Qiu, Y., Chen, Y., & Hoang, A. T. (2023). Multi-Objective Particle Swarm Optimization applied to a solar-geothermal system for electricity and hydrogen production; Utilization of zeotropic mixtures for performance improvement. *Process Safety and Environmental Protection*, *175*, 814–833. DOI: 10.1016/j.psep.2023.05.082

Zhang, Y., Hua, Q. S., Sun, L., & Liu, Q. (2020). Life cycle optimization of renewable energy systems configuration with hybrid battery/hydrogen storage: A comparative study. *Journal of Energy Storage*, *30*, 101470. DOI: 10.1016/j.est.2020.101470

Zhao, P., Gu, C., Hu, Z., Xie, D., Hernando-Gil, I., & Shen, Y. (2020). Distributionally robust hydrogen optimization with ensured security and multi-energy couplings. *IEEE Transactions on Power Systems*, *36*(1), 504–513. DOI: 10.1109/TPWRS.2020.3005991

Zheng, L., Wang, J., Yu, Y., Li, G., Zhou, M., Xia, Q., & Xu, G. (2022). On the consistency of renewable-to-hydrogen pricing. *CSEE Journal of Power and Energy Systems*, *8*(2), 392–402.

Zheng, Y., You, S., Bindner, H. W., & Münster, M. (2022). Incorporating optimal operation strategies into investment planning for wind/electrolyzer system. *CSEE Journal of Power and Energy Systems*, *8*(2), 347–359.

Zhou, X., Zhao, Q., Zhang, Y., & Sun, L. (2021). Integrated energy production unit: An innovative concept and design for energy transition toward low-carbon development. *CSEE Journal of Power and Energy Systems*, *7*(6), 1133–1139.

# Related References

To continue our tradition of advancing academic research, we have compiled a list of recommended IGI Global readings. These references will provide additional information and guidance to further enrich your knowledge and assist you with your own research and future publications.

Abbasnejad, B., Moeinzadeh, S., Ahankoob, A., & Wong, P. S. (2021). The Role of Collaboration in the Implementation of BIM-Enabled Projects. In Underwood, J., & Shelbourn, M. (Eds.), *Handbook of Research on Driving Transformational Change in the Digital Built Environment* (pp. 27–62). IGI Global. https://doi.org/10.4018/978-1-7998-6600-8.ch002

Abdulrahman, K. O., Mahamood, R. M., & Akinlabi, E. T. (2022). Additive Manufacturing (AM): Processing Technique for Lightweight Alloys and Composite Material. In K. Kumar, B. Babu, & J. Davim (Ed.), *Handbook of Research on Advancements in the Processing, Characterization, and Application of Lightweight Materials* (pp. 27-48). IGI Global. https://doi.org/10.4018/978-1-7998-7864-3.ch002

Agrawal, R., Sharma, P., & Saxena, A. (2021). A Diamond Cut Leather Substrate Antenna for BAN (Body Area Network) Application. In Singh, V., Dubey, V., Saxena, A., Tiwari, R., & Sharma, H. (Eds.), *Emerging Materials and Advanced Designs for Wearable Antennas* (pp. 54–59). IGI Global. https://doi.org/10.4018/978-1-7998-7611-3.ch004

Ahmad, F., Al-Ammar, E. A., & Alsaidan, I. (2022). Battery Swapping Station: A Potential Solution to Address the Limitations of EV Charging Infrastructure. In Alam, M., Pillai, R., & Murugesan, N. (Eds.), *Developing Charging Infrastructure and Technologies for Electric Vehicles* (pp. 195–207). IGI Global. DOI: 10.4018/978-1-7998-6858-3.ch010

Aikhuele, D. (2018). A Study of Product Development Engineering and Design Reliability Concerns. *International Journal of Applied Industrial Engineering*, *5*(1), 79–89. DOI: 10.4018/IJAIE.2018010105

Al-Khatri, H., & Al-Atrash, F. (2021). Occupants' Habits and Natural Ventilation in a Hot Arid Climate. In González-Lezcano, R. (Ed.), *Advancements in Sustainable Architecture and Energy Efficiency* (pp. 146–168). IGI Global. https://doi.org/10.4018/978-1-7998-7023-4.ch007

Al-Shebeeb, O. A., Rangaswamy, S., Gopalakrishan, B., & Devaru, D. G. (2017). Evaluation and Indexing of Process Plans Based on Electrical Demand and Energy Consumption. *International Journal of Manufacturing, Materials, and Mechanical Engineering*, *7*(3), 1–19. DOI: 10.4018/IJMMME.2017070101

Amuda, M. O., Lawal, T. F., & Akinlabi, E. T. (2017). Research Progress on Rheological Behavior of AA7075 Aluminum Alloy During Hot Deformation. *International Journal of Materials Forming and Machining Processes*, *4*(1), 53–96. DOI: 10.4018/IJMFMP.2017010104

Amuda, M. O., Lawal, T. F., & Mridha, S. (2021). Microstructure and Mechanical Properties of Silicon Carbide-Treated Ferritic Stainless Steel Welds. In Burstein, L. (Ed.), *Handbook of Research on Advancements in Manufacturing, Materials, and Mechanical Engineering* (pp. 395–411). IGI Global. https://doi.org/10.4018/978-1-7998-4939-1.ch019

Anikeev, V., Gasem, K. A., & Fan, M. (2021). Application of Supercritical Technologies in Clean Energy Production: A Review. In Chen, L. (Ed.), *Handbook of Research on Advancements in Supercritical Fluids Applications for Sustainable Energy Systems* (pp. 792–821). IGI Global. https://doi.org/10.4018/978-1-7998-5796-9.ch022

Arafat, M. Y., Saleem, I., & Devi, T. P. (2022). Drivers of EV Charging Infrastructure Entrepreneurship in India. In Alam, M., Pillai, R., & Murugesan, N. (Eds.), *Developing Charging Infrastructure and Technologies for Electric Vehicles* (pp. 208–219). IGI Global. https://doi.org/10.4018/978-1-7998-6858-3.ch011

Araujo, A., & Manninen, H. (2022). Contribution of Project-Based Learning on Social Skills Development: An Industrial Engineer Perspective. In Alves, A., & van Hattum-Janssen, N. (Eds.), *Training Engineering Students for Modern Technological Advancement* (pp. 119–145). IGI Global. https://doi.org/10.4018/978-1-7998-8816-1.ch006

Armutlu, H. (2018). Intelligent Biomedical Engineering Operations by Cloud Computing Technologies. In Kose, U., Guraksin, G., & Deperlioglu, O. (Eds.), *Nature-Inspired Intelligent Techniques for Solving Biomedical Engineering Problems* (pp. 297–317). Hershey, PA: IGI Global. DOI: 10.4018/978-1-5225-4769-3.ch015

Atik, M., Sadek, M., & Shahrour, I. (2017). Single-Run Adaptive Pushover Procedure for Shear Wall Structures. In Plevris, V., Kremmyda, G., & Fahjan, Y. (Eds.), *Performance-Based Seismic Design of Concrete Structures and Infrastructures* (pp. 59–83). Hershey, PA: IGI Global. DOI: 10.4018/978-1-5225-2089-4.ch003

Attia, H. (2021). Smart Power Microgrid Impact on Sustainable Building. In González-Lezcano, R. (Ed.), *Advancements in Sustainable Architecture and Energy Efficiency* (pp. 169–194). IGI Global. https://doi.org/10.4018/978-1-7998-7023-4.ch008

Aydin, A., Akyol, E., Gungor, M., Kaya, A., & Tasdelen, S. (2018). Geophysical Surveys in Engineering Geology Investigations With Field Examples. In Ceryan, N. (Ed.), *Handbook of Research on Trends and Digital Advances in Engineering Geology* (pp. 257–280). Hershey, PA: IGI Global. DOI: 10.4018/978-1-5225-2709-1.ch007

Ayoobkhan, M. U. D., Y., A., J., Easwaran, B., & R., T. (2021). Smart Connected Digital Products and IoT Platform With the Digital Twin. In P. Vasant, G. Weber, & W. Punurai (Ed.), *Research Advancements in Smart Technology, Optimization, and Renewable Energy* (pp. 330-350). IGI Global. https://doi.org/DOI: 10.4018/978-1-7998-3970-5.ch016

Baeza Moyano, D., & González Lezcano, R. A. (2021). The Importance of Light in Our Lives: Towards New Lighting in Schools. In González-Lezcano, R. (Ed.), *Advancements in Sustainable Architecture and Energy Efficiency* (pp. 239–256). IGI Global. https://doi.org/10.4018/978-1-7998-7023-4.ch011

Bagdadee, A. H. (2021). A Brief Assessment of the Energy Sector of Bangladesh. *International Journal of Energy Optimization and Engineering*, *10*(1), 36–55. DOI: 10.4018/IJEOE.2021010103

Baklezos, A. T., & Hadjigeorgiou, N. G. (2021). Magnetic Sensors for Space Applications and Magnetic Cleanliness Considerations. In Nikolopoulos, C. (Ed.), *Recent Trends on Electromagnetic Environmental Effects for Aeronautics and Space Applications* (pp. 147–185). IGI Global. https://doi.org/10.4018/978-1-7998-4879-0.ch006

Bas, T. G. (2017). Nutraceutical Industry with the Collaboration of Biotechnology and Nutrigenomics Engineering: The Significance of Intellectual Property in the Entrepreneurship and Scientific Research Ecosystems. In Bas, T., & Zhao, J. (Eds.), *Comparative Approaches to Biotechnology Development and Use in Developed and Emerging Nations* (pp. 1–17). Hershey, PA: IGI Global. DOI: 10.4018/978-1-5225-1040-6.ch001

Bazeer Ahamed, B., & Periakaruppan, S. (2021). Taxonomy of Influence Maximization Techniques in Unknown Social Networks. In P. Vasant, G. Weber, & W. Punurai (Eds.), *Research Advancements in Smart Technology, Optimization, and Renewable Energy* (pp. 351-363). IGI Global. https://doi.org/10.4018/978-1-7998-3970-5.ch017

Beale, R., & André, J. (2017). *Design Solutions and Innovations in Temporary Structures*. Hershey, PA: IGI Global. DOI: 10.4018/978-1-5225-2199-0

Behnam, B. (2017). Simulating Post-Earthquake Fire Loading in Conventional RC Structures. In Samui, P., Chakraborty, S., & Kim, D. (Eds.), *Modeling and Simulation Techniques in Structural Engineering* (pp. 425–444). Hershey, PA: IGI Global. DOI: 10.4018/978-1-5225-0588-4.ch015

Ben Hamida, I., Salah, S. B., Msahli, F., & Mimouni, M. F. (2018). Distribution Network Reconfiguration Using SPEA2 for Power Loss Minimization and Reliability Improvement. *International Journal of Energy Optimization and Engineering*, *7*(1), 50–65. DOI: 10.4018/IJEOE.2018010103

Bentarzi, H. (2021). Fault Tree-Based Root Cause Analysis Used to Study Mal-Operation of a Protective Relay in a Smart Grid. In Recioui, A., & Bentarzi, H. (Eds.), *Optimizing and Measuring Smart Grid Operation and Control* (pp. 289–308). IGI Global. https://doi.org/10.4018/978-1-7998-4027-5.ch012

Beysens, D. A., Garrabos, Y., & Zappoli, B. (2021). Thermal Effects in Near-Critical Fluids: Piston Effect and Related Phenomena. In Chen, L. (Ed.), *Handbook of Research on Advancements in Supercritical Fluids Applications for Sustainable Energy Systems* (pp. 1–31). IGI Global. https://doi.org/10.4018/978-1-7998-5796-9.ch001

Bhaskar, S. V., & Kudal, H. N. (2017). Effect of TiCN and AlCrN Coating on Tribological Behaviour of Plasma-nitrided AISI 4140 Steel. *International Journal of Surface Engineering and Interdisciplinary Materials Science*, *5*(2), 1–17. DOI: 10.4018/IJSEIMS.2017070101

Bhuyan, D. (2018). Designing of a Twin Tube Shock Absorber: A Study in Reverse Engineering. In Kumar, K., & Davim, J. (Eds.), *Design and Optimization of Mechanical Engineering Products* (pp. 83–104). Hershey, PA: IGI Global. DOI: 10.4018/978-1-5225-3401-3.ch005

Blumberg, G. (2021). Blockchains for Use in Construction and Engineering Projects. In Underwood, J., & Shelbourn, M. (Eds.), *Handbook of Research on Driving Transformational Change in the Digital Built Environment* (pp. 179–208). IGI Global. https://doi.org/10.4018/978-1-7998-6600-8.ch008

Bolboaca, A. M. (2021). Considerations Regarding the Use of Fuel Cells in Combined Heat and Power for Stationary Applications. In Badea, G., Felseghi, R., & A chilean, I. (Eds.), *Hydrogen Fuel Cell Technology for Stationary Applications* (pp. 239–275). IGI Global. https://doi.org/10.4018/978-1-7998-4945-2.ch010

Burstein, L. (2021). Simulation Tool for Cable Design. In Burstein, L. (Ed.), *Handbook of Research on Advancements in Manufacturing, Materials, and Mechanical Engineering* (pp. 54–74). IGI Global. https://doi.org/10.4018/978-1-7998-4939-1.ch003

Calderon, F. A., Giolo, E. G., Frau, C. D., Rengel, M. G., Rodriguez, H., Tornello, M., & Gallucci, R. (2018). Seismic Microzonation and Site Effects Detection Through Microtremors Measures: A Review. In Ceryan, N. (Ed.), *Handbook of Research on Trends and Digital Advances in Engineering Geology* (pp. 326–349). Hershey, PA: IGI Global. DOI: 10.4018/978-1-5225-2709-1.ch009

Ceryan, N., & Can, N. K. (2018). Prediction of The Uniaxial Compressive Strength of Rocks Materials. In Ceryan, N. (Ed.), *Handbook of Research on Trends and Digital Advances in Engineering Geology* (pp. 31–96). Hershey, PA: IGI Global. DOI: 10.4018/978-1-5225-2709-1.ch002

Ceryan, S. (2018). Weathering Indices Used in Evaluation of the Weathering State of Rock Material. In Ceryan, N. (Ed.), *Handbook of Research on Trends and Digital Advances in Engineering Geology* (pp. 132–186). Hershey, PA: IGI Global. DOI: 10.4018/978-1-5225-2709-1.ch004

Chen, H., Padilla, R. V., & Besarati, S. (2017). Supercritical Fluids and Their Applications in Power Generation. In Chen, L., & Iwamoto, Y. (Eds.), *Advanced Applications of Supercritical Fluids in Energy Systems* (pp. 369–402). Hershey, PA: IGI Global. DOI: 10.4018/978-1-5225-2047-4.ch012

Chen, H., Padilla, R. V., & Besarati, S. (2021). Supercritical Fluids and Their Applications in Power Generation. In Chen, L. (Ed.), *Handbook of Research on Advancements in Supercritical Fluids Applications for Sustainable Energy Systems* (pp. 566–599). IGI Global. https://doi.org/10.4018/978-1-7998-5796-9.ch016

Chen, L. (2017). Principles, Experiments, and Numerical Studies of Supercritical Fluid Natural Circulation System. In Chen, L., & Iwamoto, Y. (Eds.), *Advanced Applications of Supercritical Fluids in Energy Systems* (pp. 136–187). Hershey, PA: IGI Global. DOI: 10.4018/978-1-5225-2047-4.ch005

Chen, L. (2021). Principles, Experiments, and Numerical Studies of Supercritical Fluid Natural Circulation System. In Chen, L. (Ed.), *Handbook of Research on Advancements in Supercritical Fluids Applications for Sustainable Energy Systems* (pp. 219–269). IGI Global. https://doi.org/10.4018/978-1-7998-5796-9.ch007

Chiba, Y., Marif, Y., Henini, N., & Tlemcani, A. (2021). Modeling of Magnetic Refrigeration Device by Using Artificial Neural Networks Approach. *International Journal of Energy Optimization and Engineering, 10*(4), 68–76. https://doi.org/10.4018/IJEOE.2021100105

Clementi, F., Di Sciascio, G., Di Sciascio, S., & Lenci, S. (2017). Influence of the Shear-Bending Interaction on the Global Capacity of Reinforced Concrete Frames: A Brief Overview of the New Perspectives. In Plevris, V., Kremmyda, G., & Fahjan, Y. (Eds.), *Performance-Based Seismic Design of Concrete Structures and Infrastructures* (pp. 84–111). Hershey, PA: IGI Global. DOI: 10.4018/978-1-5225-2089-4.ch004

Codinhoto, R., Fialho, B. C., Pinti, L., & Fabricio, M. M. (2021). BIM and IoT for Facilities Management: Understanding Key Maintenance Issues. In Underwood, J., & Shelbourn, M. (Eds.), *Handbook of Research on Driving Transformational Change in the Digital Built Environment* (pp. 209–231). IGI Global. DOI: 10.4018/978-1-7998-6600-8.ch009

Cortés-Polo, D., Calle-Cancho, J., Carmona-Murillo, J., & González-Sánchez, J. (2017). Future Trends in Mobile-Fixed Integration for Next Generation Networks: Classification and Analysis. *International Journal of Vehicular Telematics and Infotainment Systems, 1*(1), 33–53. DOI: 10.4018/IJVTIS.2017010103

Costa, H. G., Sheremetieff, F. H., & Araújo, E. A. (2022). Influence of Game-Based Methods in Developing Engineering Competences. In Alves, A., & van Hattum-Janssen, N. (Eds.), *Training Engineering Students for Modern Technological Advancement* (pp. 69–88). IGI Global. https://doi.org/10.4018/978-1-7998-8816-1.ch004

Cui, X., Zeng, S., Li, Z., Zheng, Q., Yu, X., & Han, B. (2018). Advanced Composites for Civil Engineering Infrastructures. In Kumar, K., & Davim, J. (Eds.), *Composites and Advanced Materials for Industrial Applications* (pp. 212–248). Hershey, PA: IGI Global. DOI: 10.4018/978-1-5225-5216-1.ch010

Dalgıç, S., & Kuşku, İ. (2018). Geological and Geotechnical Investigations in Tunneling. In Ceryan, N. (Ed.), *Handbook of Research on Trends and Digital Advances in Engineering Geology* (pp. 482–529). Hershey, PA: IGI Global. DOI: 10.4018/978-1-5225-2709-1.ch014

Dang, C., & Hihara, E. (2021). Study on Cooling Heat Transfer of Supercritical Carbon Dioxide Applied to Transcritical Carbon Dioxide Heat Pump. In Chen, L. (Ed.), *Handbook of Research on Advancements in Supercritical Fluids Applications for Sustainable Energy Systems* (pp. 451–493). IGI Global. https://doi.org/10.4018/978-1-7998-5796-9.ch013

Daus, Y., Kharchenko, V., & Yudaev, I. (2021). Research of Solar Energy Potential of Photovoltaic Installations on Enclosing Structures of Buildings. *International Journal of Energy Optimization and Engineering*, *10*(4), 18–34. https://doi.org/10.4018/IJEOE.2021100102

Daus, Y., Kharchenko, V., & Yudaev, I. (2021). Optimizing Layout of Distributed Generation Sources of Power Supply System of Agricultural Object. *International Journal of Energy Optimization and Engineering*, *10*(3), 70–84. https://doi.org/10.4018/IJEOE.2021070104

de la Varga, D., Soto, M., Arias, C. A., van Oirschot, D., Kilian, R., Pascual, A., & Álvarez, J. A. (2017). Constructed Wetlands for Industrial Wastewater Treatment and Removal of Nutrients. In Val del Río, Á., Campos Gómez, J., & Mosquera Corral, A. (Eds.), *Technologies for the Treatment and Recovery of Nutrients from Industrial Wastewater* (pp. 202–230). Hershey, PA: IGI Global. DOI: 10.4018/978-1-5225-1037-6.ch008

Deb, S., Ammar, E. A., AlRajhi, H., Alsaidan, I., & Shariff, S. M. (2022). V2G Pilot Projects: Review and Lessons Learnt. In Alam, M., Pillai, R., & Murugesan, N. (Eds.), *Developing Charging Infrastructure and Technologies for Electric Vehicles* (pp. 252–267). IGI Global. https://doi.org/10.4018/978-1-7998-6858-3.ch014

Dekhandji, F. Z., & Rais, M. C. (2021). A Comparative Study of Power Quality Monitoring Using Various Techniques. In Recioui, A., & Bentarzi, H. (Eds.), *Optimizing and Measuring Smart Grid Operation and Control* (pp. 259–288). IGI Global. https://doi.org/10.4018/978-1-7998-4027-5.ch011

Deperlioglu, O. (2018). Intelligent Techniques Inspired by Nature and Used in Biomedical Engineering. In Kose, U., Guraksin, G., & Deperlioglu, O. (Eds.), *Nature-Inspired Intelligent Techniques for Solving Biomedical Engineering Problems* (pp. 51–77). Hershey, PA: IGI Global. DOI: 10.4018/978-1-5225-4769-3.ch003

Dhurpate, P. R., & Tang, H. (2021). Quantitative Analysis of the Impact of Inter-Line Conveyor Capacity for Throughput of Manufacturing Systems. *International Journal of Manufacturing, Materials, and Mechanical Engineering, 11*(1), 1–17. https://doi.org/10.4018/IJMMME.2021010101

Dinkar, S., & Deep, K. (2021). A Survey of Recent Variants and Applications of Antlion Optimizer. *International Journal of Energy Optimization and Engineering, 10*(2), 48–73. DOI: 10.4018/IJEOE.2021040103

Dixit, A. (2018). Application of Silica-Gel-Reinforced Aluminium Composite on the Piston of Internal Combustion Engine: Comparative Study of Silica-Gel-Reinforced Aluminium Composite Piston With Aluminium Alloy Piston. In Kumar, K., & Davim, J. (Eds.), *Composites and Advanced Materials for Industrial Applications* (pp. 63–98). Hershey, PA: IGI Global. DOI: 10.4018/978-1-5225-5216-1.ch004

Drabecki, M. P., & Kułak, K. B. (2021). Global Pandemics on European Electrical Energy Markets: Lessons Learned From the COVID-19 Outbreak. *International Journal of Energy Optimization and Engineering, 10*(3), 24–46. https://doi.org/10.4018/IJEOE.2021070102

Dutta, M. M. (2021). Nanomaterials for Food and Agriculture. In Bhat, M., Wani, I., & Ashraf, S. (Eds.), *Applications of Nanomaterials in Agriculture, Food Science, and Medicine* (pp. 75–97). IGI Global. DOI: 10.4018/978-1-7998-5563-7.ch004

Dutta, M. M., & Goswami, M. (2021). Coating Materials: Nano-Materials. In Roy, S., & Bose, G. (Eds.), *Advanced Surface Coating Techniques for Modern Industrial Applications* (pp. 1–30). IGI Global. DOI: 10.4018/978-1-7998-4870-7.ch001

Elsayed, A. M., Dakkama, H. J., Mahmoud, S., Al-Dadah, R., & Kaialy, W. (2017). Sustainable Cooling Research Using Activated Carbon Adsorbents and Their Environmental Impact. In Kobayashi, T. (Ed.), *Applied Environmental Materials Science for Sustainability* (pp. 186–221). Hershey, PA: IGI Global. DOI: 10.4018/978-1-5225-1971-3.ch009

Ercanoglu, M., & Sonmez, H. (2018). General Trends and New Perspectives on Landslide Mapping and Assessment Methods. In Ceryan, N. (Ed.), *Handbook of Research on Trends and Digital Advances in Engineering Geology* (pp. 350–379). Hershey, PA: IGI Global. DOI: 10.4018/978-1-5225-2709-1.ch010

Faroz, S. A., Pujari, N. N., Rastogi, R., & Ghosh, S. (2017). Risk Analysis of Structural Engineering Systems Using Bayesian Inference. In Samui, P., Chakraborty, S., & Kim, D. (Eds.), *Modeling and Simulation Techniques in Structural Engineering* (pp. 390–424). Hershey, PA: IGI Global. DOI: 10.4018/978-1-5225-0588-4.ch014

Fekik, A., Hamida, M. L., Denoun, H., Azar, A. T., Kamal, N. A., Vaidyanathan, S., Bousbaine, A., & Benamrouche, N. (2022). Multilevel Inverter for Hybrid Fuel Cell/PV Energy Conversion System. In Fekik, A., & Benamrouche, N. (Eds.), *Modeling and Control of Static Converters for Hybrid Storage Systems* (pp. 233–270). IGI Global. https://doi.org/10.4018/978-1-7998-7447-8.ch009

Fekik, A., Hamida, M. L., Houassine, H., Azar, A. T., Kamal, N. A., Denoun, H., Vaidyanathan, S., & Sambas, A. (2022). Power Quality Improvement for Grid-Connected Photovoltaic Panels Using Direct Power Control. In Fekik, A., & Benamrouche, N. (Eds.), *Modeling and Control of Static Converters for Hybrid Storage Systems* (pp. 107–142). IGI Global. https://doi.org/10.4018/978-1-7998-7447-8.ch005

Fernando, P. R., Hamigah, T., Disne, S., Wickramasingha, G. G., & Sutharshan, A. (2018). The Evaluation of Engineering Properties of Low Cost Concrete Blocks by Partial Doping of Sand with Sawdust: Low Cost Sawdust Concrete Block. *International Journal of Strategic Engineering*, *1*(2), 26–42. DOI: 10.4018/IJoSE.2018070103

Ferro, G., Minciardi, R., Parodi, L., & Robba, M. (2022). Optimal Charging Management of Microgrid-Integrated Electric Vehicles. In Alam, M., Pillai, R., & Murugesan, N. (Eds.), *Developing Charging Infrastructure and Technologies for Electric Vehicles* (pp. 133–155). IGI Global. https://doi.org/10.4018/978-1-7998-6858-3.ch007

Flumerfelt, S., & Green, C. (2022). Graduate Lean Leadership Education: A Case Study of a Program. In Alves, A., & van Hattum-Janssen, N. (Eds.), *Training Engineering Students for Modern Technological Advancement* (pp. 202–224). IGI Global. https://doi.org/10.4018/978-1-7998-8816-1.ch010

Galli, B. J. (2021). Implications of Economic Decision Making to the Project Manager. *International Journal of Strategic Engineering*, *4*(1), 19–32. https://doi.org/10.4018/IJoSE.2021010102

Gento, A. M., Pimentel, C., & Pascual, J. A. (2022). Teaching Circular Economy and Lean Management in a Learning Factory. In Alves, A., & van Hattum-Janssen, N. (Eds.), *Training Engineering Students for Modern Technological Advancement* (pp. 183–201). IGI Global. https://doi.org/10.4018/978-1-7998-8816-1.ch009

Ghosh, S., Mitra, S., Ghosh, S., & Chakraborty, S. (2017). Seismic Reliability Analysis in the Framework of Metamodelling Based Monte Carlo Simulation. In Samui, P., Chakraborty, S., & Kim, D. (Eds.), *Modeling and Simulation Techniques in Structural Engineering* (pp. 192–208). Hershey, PA: IGI Global. DOI: 10.4018/978-1-5225-0588-4.ch006

Gil, M., & Otero, B. (2017). Learning Engineering Skills through Creativity and Collaboration: A Game-Based Proposal. In Alexandre Peixoto de Queirós, R., & Pinto, M. (Eds.), *Gamification-Based E-Learning Strategies for Computer Programming Education* (pp. 14–29). Hershey, PA: IGI Global. DOI: 10.4018/978-1-5225-1034-5.ch002

Gill, J., Ayre, M., & Mills, J. (2017). Revisioning the Engineering Profession: How to Make It Happen! In Gray, M., & Thomas, K. (Eds.), *Strategies for Increasing Diversity in Engineering Majors and Careers* (pp. 156–175). Hershey, PA: IGI Global. DOI: 10.4018/978-1-5225-2212-6.ch008

Godzhaev, Z., Senkevich, S., Kuzmin, V., & Melikov, I. (2021). Use of the Neural Network Controller of Sprung Mass to Reduce Vibrations From Road Irregularities. In P. Vasant, G. Weber, & W. Punurai (Ed.), *Research Advancements in Smart Technology, Optimization, and Renewable Energy* (pp. 69-87). IGI Global. https://doi.org/10.4018/978-1-7998-3970-5.ch005

Gomes de Gusmão, C. M. (2022). Digital Competencies and Transformation in Higher Education: Upskilling With Extension Actions. In Alves, A., & van Hattum-Janssen, N. (Eds.), *Training Engineering Students for Modern Technological Advancement* (pp. 313–328). IGI Global. https://doi.org/10.4018/978-1-7998-8816-1.ch015A

Goyal, N., Ram, M., & Kumar, P. (2017). Welding Process under Fault Coverage Approach for Reliability and MTTF. In Ram, M., & Davim, J. (Eds.), *Mathematical Concepts and Applications in Mechanical Engineering and Mechatronics* (pp. 222–245). Hershey, PA: IGI Global. DOI: 10.4018/978-1-5225-1639-2.ch011

Gray, M., & Lundy, C. (2017). Engineering Study Abroad: High Impact Strategy for Increasing Access. In Gray, M., & Thomas, K. (Eds.), *Strategies for Increasing Diversity in Engineering Majors and Careers* (pp. 42–59). Hershey, PA: IGI Global. DOI: 10.4018/978-1-5225-2212-6.ch003

Güler, O., & Varol, T. (2021). Fabrication of Functionally Graded Metal and Ceramic Powders Synthesized by Electroless Deposition. In Roy, S., & Bose, G. (Eds.), *Advanced Surface Coating Techniques for Modern Industrial Applications* (pp. 150–187). IGI Global. https://doi.org/10.4018/978-1-7998-4870-7.ch007

Guraksin, G. E. (2018). Internet of Things and Nature-Inspired Intelligent Techniques for the Future of Biomedical Engineering. In Kose, U., Guraksin, G., & Deperlioglu, O. (Eds.), *Nature-Inspired Intelligent Techniques for Solving Biomedical Engineering Problems* (pp. 263–282). Hershey, PA: IGI Global. DOI: 10.4018/978-1-5225-4769-3.ch013

Hamida, M. L., Fekik, A., Denoun, H., Ardjal, A., & Bokhtache, A. A. (2022). Flying Capacitor Inverter Integration in a Renewable Energy System. In Fekik, A., & Benamrouche, N. (Eds.), *Modeling and Control of Static Converters for Hybrid Storage Systems* (pp. 287–306). IGI Global. https://doi.org/10.4018/978-1-7998-7447-8.ch011

Hasegawa, N., & Takahashi, Y. (2021). Control of Soap Bubble Ejection Robot Using Facial Expressions. *International Journal of Manufacturing, Materials, and Mechanical Engineering, 11*(2), 1–16. https://doi.org/10.4018/IJMMME.2021040101

Hejazi, T., & Akbari, L. (2017). A Multiresponse Optimization Model for Statistical Design of Processes with Discrete Variables. In Ram, M., & Davim, J. (Eds.), *Mathematical Concepts and Applications in Mechanical Engineering and Mechatronics* (pp. 17–37). Hershey, PA: IGI Global. DOI: 10.4018/978-1-5225-1639-2.ch002

Hejazi, T., & Hejazi, A. (2017). Monte Carlo Simulation for Reliability-Based Design of Automotive Complex Subsystems. In Ram, M., & Davim, J. (Eds.), *Mathematical Concepts and Applications in Mechanical Engineering and Mechatronics* (pp. 177–200). Hershey, PA: IGI Global. DOI: 10.4018/978-1-5225-1639-2.ch009

Hejazi, T., & Poursabbagh, H. (2017). Reliability Analysis of Engineering Systems: An Accelerated Life Testing for Boiler Tubes. In Ram, M., & Davim, J. (Eds.), *Mathematical Concepts and Applications in Mechanical Engineering and Mechatronics* (pp. 154–176). Hershey, PA: IGI Global. DOI: 10.4018/978-1-5225-1639-2.ch008

Henao, J., Poblano-Salas, C. A., Vargas, F., Giraldo-Betancur, A. L., Corona-Castuera, J., & Sotelo-Mazón, O. (2021). Principles and Applications of Thermal Spray Coatings. In Roy, S., & Bose, G. (Eds.), *Advanced Surface Coating Techniques for Modern Industrial Applications* (pp. 31–70). IGI Global. https://doi.org/10.4018/978-1-7998-4870-7.ch002

Henao, J., & Sotelo, O. (2018). Surface Engineering at High Temperature: Thermal Cycling and Corrosion Resistance. In Pakseresht, A. (Ed.), *Production, Properties, and Applications of High Temperature Coatings* (pp. 131–159). Hershey, PA: IGI Global. DOI: 10.4018/978-1-5225-4194-3.ch006

Hrnčič, M. K., Cör, D., & Knez, Ž. (2021). Supercritical Fluids as a Tool for Green Energy and Chemicals. In Chen, L. (Ed.), *Handbook of Research on Advancements in Supercritical Fluids Applications for Sustainable Energy Systems* (pp. 761–791). IGI Global. DOI: 10.4018/978-1-7998-5796-9.ch021

Ibrahim, O., Erdem, S., & Gurbuz, E. (2021). Studying Physical and Chemical Properties of Graphene Oxide and Reduced Graphene Oxide and Their Applications in Sustainable Building Materials. In González-Lezcano, R. (Ed.), *Advancements in Sustainable Architecture and Energy Efficiency* (pp. 221–238). IGI Global. https://doi.org/10.4018/978-1-7998-7023-4.ch010

Ihianle, I. K., Islam, S., Naeem, U., & Ebenuwa, S. H. (2021). Exploiting Patterns of Object Use for Human Activity Recognition. In Nwajana, A., & Ihianle, I. (Eds.), *Handbook of Research on 5G Networks and Advancements in Computing, Electronics, and Electrical Engineering* (pp. 382–401). IGI Global. https://doi.org/10.4018/978-1-7998-6992-4.ch015

Ijemaru, G. K., Ngharamike, E. T., Oleka, E. U., & Nwajana, A. O. (2021). An Energy-Efficient Model for Opportunistic Data Collection in IoV-Enabled SC Waste Management. In Nwajana, A., & Ihianle, I. (Eds.), *Handbook of Research on 5G Networks and Advancements in Computing, Electronics, and Electrical Engineering* (pp. 1–19). IGI Global. https://doi.org/10.4018/978-1-7998-6992-4.ch001

Ilori, O. O., Adetan, D. A., & Umoru, L. E. (2017). Effect of Cutting Parameters on the Surface Residual Stress of Face-Milled Pearlitic Ductile Iron. *International Journal of Materials Forming and Machining Processes*, *4*(1), 38–52. DOI: 10.4018/IJMFMP.2017010103

Imam, M. H., Tasadduq, I. A., Ahmad, A., Aldosari, F., & Khan, H. (2017). Automated Generation of Course Improvement Plans Using Expert System. *International Journal of Quality Assurance in Engineering and Technology Education*, *6*(1), 1–12. DOI: 10.4018/IJQAETE.2017010101

Injeti, S. K., & Kumar, T. V. (2018). A WDO Framework for Optimal Deployment of DGs and DSCs in a Radial Distribution System Under Daily Load Pattern to Improve Techno-Economic Benefits. *International Journal of Energy Optimization and Engineering*, *7*(2), 1–38. DOI: 10.4018/IJEOE.2018040101

Ishii, N., Anami, K., & Knisely, C. W. (2018). *Dynamic Stability of Hydraulic Gates and Engineering for Flood Prevention*. Hershey, PA: IGI Global. DOI: 10.4018/978-1-5225-3079-4

Iwamoto, Y., & Yamaguchi, H. (2021). Application of Supercritical Carbon Dioxide for Solar Water Heater. In Chen, L. (Ed.), *Handbook of Research on Advancements in Supercritical Fluids Applications for Sustainable Energy Systems* (pp. 370–387). IGI Global. https://doi.org/10.4018/978-1-7998-5796-9.ch010

Jayapalan, S. (2018). A Review of Chemical Treatments on Natural Fibers-Based Hybrid Composites for Engineering Applications. In Kumar, K., & Davim, J. (Eds.), *Composites and Advanced Materials for Industrial Applications* (pp. 16–37). Hershey, PA: IGI Global. DOI: 10.4018/978-1-5225-5216-1.ch002

Kapetanakis, T. N., Vardiambasis, I. O., Ioannidou, M. P., & Konstantaras, A. I. (2021). Modeling Antenna Radiation Using Artificial Intelligence Techniques: The Case of a Circular Loop Antenna. In Nikolopoulos, C. (Ed.), *Recent Trends on Electromagnetic Environmental Effects for Aeronautics and Space Applications* (pp. 186–225). IGI Global. https://doi.org/10.4018/978-1-7998-4879-0.ch007

Karkalos, N. E., Markopoulos, A. P., & Dossis, M. F. (2017). Optimal Model Parameters of Inverse Kinematics Solution of a 3R Robotic Manipulator Using ANN Models. *International Journal of Manufacturing, Materials, and Mechanical Engineering*, *7*(3), 20–40. DOI: 10.4018/IJMMME.2017070102

Kelly, M., Costello, M., Nicholson, G., & O'Connor, J. (2021). The Evolving Integration of BIM Into Built Environment Programmes in a Higher Education Institute. In Underwood, J., & Shelbourn, M. (Eds.), *Handbook of Research on Driving Transformational Change in the Digital Built Environment* (pp. 294–326). IGI Global. https://doi.org/10.4018/978-1-7998-6600-8.ch012

Kesimal, A., Karaman, K., Cihangir, F., & Ercikdi, B. (2018). Excavatability Assessment of Rock Masses for Geotechnical Studies. In Ceryan, N. (Ed.), *Handbook of Research on Trends and Digital Advances in Engineering Geology* (pp. 231–256). Hershey, PA: IGI Global. DOI: 10.4018/978-1-5225-2709-1.ch006

Knoflacher, H. (2017). The Role of Engineers and Their Tools in the Transport Sector after Paradigm Change: From Assumptions and Extrapolations to Science. In Knoflacher, H., & Ocalir-Akunal, E. (Eds.), *Engineering Tools and Solutions for Sustainable Transportation Planning* (pp. 1–29). Hershey, PA: IGI Global. DOI: 10.4018/978-1-5225-2116-7.ch001

Kose, U. (2018). Towards an Intelligent Biomedical Engineering With Nature-Inspired Artificial Intelligence Techniques. In Kose, U., Guraksin, G., & Deperlioglu, O. (Eds.), *Nature-Inspired Intelligent Techniques for Solving Biomedical Engineering Problems* (pp. 1–26). Hershey, PA: IGI Global. DOI: 10.4018/978-1-5225-4769-3.ch001

Kostić, S. (2018). A Review on Enhanced Stability Analyses of Soil Slopes Using Statistical Design. In Ceryan, N. (Ed.), *Handbook of Research on Trends and Digital Advances in Engineering Geology* (pp. 446–481). Hershey, PA: IGI Global. DOI: 10.4018/978-1-5225-2709-1.ch013

Kumar, A., Patil, P. P., & Prajapati, Y. K. (2018). *Advanced Numerical Simulations in Mechanical Engineering*. Hershey, PA: IGI Global. DOI: 10.4018/978-1-5225-3722-9

Kumar, G. R., Rajyalakshmi, G., & Manupati, V. K. (2017). Surface Micro Patterning of Aluminium Reinforced Composite through Laser Peening. *International Journal of Manufacturing, Materials, and Mechanical Engineering*, *7*(4), 15–27. DOI: 10.4018/IJMMME.2017100102

Kumar, N., Basu, D. N., & Chen, L. (2021). Effect of Flow Acceleration and Buoyancy on Thermalhydraulics of sCO2 in Mini/Micro-Channel. In Chen, L. (Ed.), *Handbook of Research on Advancements in Supercritical Fluids Applications for Sustainable Energy Systems* (pp. 161–182). IGI Global. DOI: 10.4018/978-1-7998-5796-9.ch005

Kumari, N., & Kumar, K. (2018). Fabrication of Orthotic Calipers With Epoxy-Based Green Composite. In Kumar, K., & Davim, J. (Eds.), *Composites and Advanced Materials for Industrial Applications* (pp. 157–176). Hershey, PA: IGI Global. DOI: 10.4018/978-1-5225-5216-1.ch008

Kuppusamy, R. R. (2018). Development of Aerospace Composite Structures Through Vacuum-Enhanced Resin Transfer Moulding Technology (VERTMTy): Vacuum-Enhanced Resin Transfer Moulding. In Kumar, K., & Davim, J. (Eds.), *Composites and Advanced Materials for Industrial Applications* (pp. 99–111). Hershey, PA: IGI Global. DOI: 10.4018/978-1-5225-5216-1.ch005

Kurganov, V. A., Zeigarnik, Y. A., & Maslakova, I. V. (2021). Normal and Deteriorated Heat Transfer Under Heating Turbulent Supercritical Pressure Coolants Flows in Round Tubes. In Chen, L. (Ed.), *Handbook of Research on Advancements in Supercritical Fluids Applications for Sustainable Energy Systems* (pp. 494–532). IGI Global. https://doi.org/10.4018/978-1-7998-5796-9.ch014

Li, H., & Zhang, Y. (2021). Heat Transfer and Fluid Flow Modeling for Supercritical Fluids in Advanced Energy Systems. In Chen, L. (Ed.), *Handbook of Research on Advancements in Supercritical Fluids Applications for Sustainable Energy Systems* (pp. 388–422). IGI Global. https://doi.org/10.4018/978-1-7998-5796-9.ch011

Loy, J., Howell, S., & Cooper, R. (2017). Engineering Teams: Supporting Diversity in Engineering Education. In Gray, M., & Thomas, K. (Eds.), *Strategies for Increasing Diversity in Engineering Majors and Careers* (pp. 106–129). Hershey, PA: IGI Global. DOI: 10.4018/978-1-5225-2212-6.ch006

Macher, G., Armengaud, E., Kreiner, C., Brenner, E., Schmittner, C., Ma, Z., & Krammer, M. (2018). Integration of Security in the Development Lifecycle of Dependable Automotive CPS. In Druml, N., Genser, A., Krieg, A., Menghin, M., & Hoeller, A. (Eds.), *Solutions for Cyber-Physical Systems Ubiquity* (pp. 383–423). Hershey, PA: IGI Global. DOI: 10.4018/978-1-5225-2845-6.ch015

Madhu, M. N., Singh, J. G., Mohan, V., & Ongsakul, W. (2021). Transmission Risk Optimization in Interconnected Systems: Risk-Adjusted Available Transfer Capability. In P. Vasant, G. Weber, & W. Punurai (Ed.), *Research Advancements in Smart Technology, Optimization, and Renewable Energy* (pp. 183-199). IGI Global. https://doi.org/10.4018/978-1-7998-3970-5.ch010

Mahendramani, G., & Lakshmana Swamy, N. (2018). Effect of Weld Groove Area on Distortion of Butt Welded Joints in Submerged Arc Welding. *International Journal of Manufacturing, Materials, and Mechanical Engineering*, *8*(2), 33–44. DOI: 10.4018/IJMMME.2018040103

Makropoulos, G., Koumaras, H., Setaki, F., Filis, K., Lutz, T., Montowtt, P., Tomaszewski, L., Dybiec, P., & Järvet, T. (2021). 5G and Unmanned Aerial Vehicles (UAVs) Use Cases: Analysis of the Ecosystem, Architecture, and Applications. In Nwajana, A., & Ihianle, I. (Eds.), *Handbook of Research on 5G Networks and Advancements in Computing, Electronics, and Electrical Engineering* (pp. 36–69). IGI Global. https://doi.org/10.4018/978-1-7998-6992-4.ch003

Meric, E. M., Erdem, S., & Gurbuz, E. (2021). Application of Phase Change Materials in Construction Materials for Thermal Energy Storage Systems in Buildings. In González-Lezcano, R. (Ed.), *Advancements in Sustainable Architecture and Energy Efficiency* (pp. 1–20). IGI Global. https://doi.org/10.4018/978-1-7998-7023-4.ch001

Mihret, E. T., & Yitayih, K. A. (2021). Operation of VANET Communications: The Convergence of UAV System With LTE/4G and WAVE Technologies. *International Journal of Smart Vehicles and Smart Transportation, 4*(1), 29–51. https://doi.org/10.4018/IJSVST.2021010103

Mir, M. A., Bhat, B. A., Sheikh, B. A., Rather, G. A., Mehraj, S., & Mir, W. R. (2021). Nanomedicine in Human Health Therapeutics and Drug Delivery: Nanobiotechnology and Nanobiomedicine. In Bhat, M., Wani, I., & Ashraf, S. (Eds.), *Applications of Nanomaterials in Agriculture, Food Science, and Medicine* (pp. 229–251). IGI Global. DOI: 10.4018/978-1-7998-5563-7.ch013

Mohammadzadeh, S., & Kim, Y. (2017). Nonlinear System Identification of Smart Buildings. In Samui, P., Chakraborty, S., & Kim, D. (Eds.), *Modeling and Simulation Techniques in Structural Engineering* (pp. 328–347). Hershey, PA: IGI Global. DOI: 10.4018/978-1-5225-0588-4.ch011

Molina, G. J., Aktaruzzaman, F., Soloiu, V., & Rahman, M. (2017). Design and Testing of a Jet-Impingement Instrument to Study Surface-Modification Effects by Nanofluids. *International Journal of Surface Engineering and Interdisciplinary Materials Science, 5*(2), 43–61. DOI: 10.4018/IJSEIMS.2017070104

Moreno-Rangel, A., & Carrillo, G. (2021). Energy-Efficient Homes: A Heaven for Respiratory Illnesses. In González-Lezcano, R. (Ed.), *Advancements in Sustainable Architecture and Energy Efficiency* (pp. 49–71). IGI Global. https://doi.org/10.4018/978-1-7998-7023-4.ch003

Msomi, V., & Jantjies, B. T. (2021). Correlative Analysis Between Tensile Properties and Tool Rotational Speeds of Friction Stir Welded Similar Aluminium Alloy Joints. *International Journal of Surface Engineering and Interdisciplinary Materials Science, 9*(2), 58–78. https://doi.org/10.4018/IJSEIMS.2021070104

Muigai, M. N., Mwema, F. M., Akinlabi, E. T., & Obiko, J. O. (2021). Surface Engineering of Materials Through Weld-Based Technologies: An Overview. In Roy, S., & Bose, G. (Eds.), *Advanced Surface Coating Techniques for Modern Industrial Applications* (pp. 247–260). IGI Global. DOI: 10.4018/978-1-7998-4870-7.ch011

Mukherjee, A., Saeed, R. A., Dutta, S., & Naskar, M. K. (2017). Fault Tracking Framework for Software-Defined Networking (SDN). In Singhal, C., & De, S. (Eds.), *Resource Allocation in Next-Generation Broadband Wireless Access Networks* (pp. 247–272). Hershey, PA: IGI Global. DOI: 10.4018/978-1-5225-2023-8.ch011

Mukhopadhyay, A., Barman, T. K., & Sahoo, P. (2018). Electroless Nickel Coatings for High Temperature Applications. In Kumar, K., & Davim, J. (Eds.), *Composites and Advanced Materials for Industrial Applications* (pp. 297–331). Hershey, PA: IGI Global. DOI: 10.4018/978-1-5225-5216-1.ch013

Mwema, F. M., & Wambua, J. M. (2022). Machining of Poly Methyl Methacrylate (PMMA) and Other Olymeric Materials: A Review. In Kumar, K., Babu, B., & Davim, J. (Eds.), *Handbook of Research on Advancements in the Processing, Characterization, and Application of Lightweight Materials* (pp. 363–379). IGI Global. https://doi.org/10.4018/978-1-7998-7864-3.ch016

Mykhailyshyn, R., Savkiv, V., Boyko, I., Prada, E., & Virgala, I. (2021). Substantiation of Parameters of Friction Elements of Bernoulli Grippers With a Cylindrical Nozzle. *International Journal of Manufacturing, Materials, and Mechanical Engineering, 11*(2), 17–39. https://doi.org/10.4018/IJMMME.2021040102

Náprstek, J., & Fischer, C. (2017). Dynamic Stability and Post-Critical Processes of Slender Auto-Parametric Systems. In Plevris, V., Kremmyda, G., & Fahjan, Y. (Eds.), *Performance-Based Seismic Design of Concrete Structures and Infrastructures* (pp. 128–171). Hershey, PA: IGI Global. DOI: 10.4018/978-1-5225-2089-4.ch006

Nautiyal, L., Shivach, P., & Ram, M. (2018). Optimal Designs by Means of Genetic Algorithms. In Ram, M., & Davim, J. (Eds.), *Soft Computing Techniques and Applications in Mechanical Engineering* (pp. 151–161). Hershey, PA: IGI Global. DOI: 10.4018/978-1-5225-3035-0.ch007

Nazir, R. (2017). Advanced Nanomaterials for Water Engineering and Treatment: Nano-Metal Oxides and Their Nanocomposites. In Saleh, T. (Ed.), *Advanced Nanomaterials for Water Engineering, Treatment, and Hydraulics* (pp. 84–126). Hershey, PA: IGI Global. DOI: 10.4018/978-1-5225-2136-5.ch005

Nikolopoulos, C. D. (2021). Recent Advances on Measuring and Modeling ELF-Radiated Emissions for Space Applications. In Nikolopoulos, C. (Ed.), *Recent Trends on Electromagnetic Environmental Effects for Aeronautics and Space Applications* (pp. 1–38). IGI Global. https://doi.org/10.4018/978-1-7998-4879-0.ch001

Nogueira, A. F., Ribeiro, J. C., Fernández de Vega, F., & Zenha-Rela, M. A. (2018). Evolutionary Approaches to Test Data Generation for Object-Oriented Software: Overview of Techniques and Tools. In M. Khosrow-Pour, D.B.A. (Ed.), *Incorporating Nature-Inspired Paradigms in Computational Applications* (pp. 162-194). Hershey, PA: IGI Global. https://doi.org/DOI: 10.4018/978-1-5225-5020-4.ch006

Nwajana, A. O., Obi, E. R., Ijemaru, G. K., Oleka, E. U., & Anthony, D. C. (2021). Fundamentals of RF/Microwave Bandpass Filter Design. In Nwajana, A., & Ihianle, I. (Eds.), *Handbook of Research on 5G Networks and Advancements in Computing, Electronics, and Electrical Engineering* (pp. 149–164). IGI Global. https://doi.org/10.4018/978-1-7998-6992-4.ch005

Ogbodo, E. A. (2021). Comparative Study of Transmission Line Junction vs. Asynchronously Coupled Junction Diplexers. In Nwajana, A., & Ihianle, I. (Eds.), *Handbook of Research on 5G Networks and Advancements in Computing, Electronics, and Electrical Engineering* (pp. 326–336). IGI Global. https://doi.org/10.4018/978-1-7998-6992-4.ch013

Orosa, J. A., Vergara, D., Fraguela, F., & Masdías-Bonome, A. (2021). Statistical Understanding and Optimization of Building Energy Consumption and Climate Change Consequences. In González-Lezcano, R. (Ed.), *Advancements in Sustainable Architecture and Energy Efficiency* (pp. 195–220). IGI Global. https://doi.org/10.4018/978-1-7998-7023-4.ch009

Osho, M. B. (2018). Industrial Enzyme Technology: Potential Applications. In Bharati, S., & Chaurasia, P. (Eds.), *Research Advancements in Pharmaceutical, Nutritional, and Industrial Enzymology* (pp. 375–394). Hershey, PA: IGI Global. DOI: 10.4018/978-1-5225-5237-6.ch017

Ouadi, A., & Zitouni, A. (2021). Phasor Measurement Improvement Using Digital Filter in a Smart Grid. In Recioui, A., & Bentarzi, H. (Eds.), *Optimizing and Measuring Smart Grid Operation and Control* (pp. 100–117). IGI Global. https://doi.org/10.4018/978-1-7998-4027-5.ch005

Padmaja, P., & Marutheswar, G. (2017). Certain Investigation on Secured Data Transmission in Wireless Sensor Networks. *International Journal of Mobile Computing and Multimedia Communications*, *8*(1), 48–61. DOI: 10.4018/IJMCMC.2017010104

Palmer, S., & Hall, W. (2017). An Evaluation of Group Work in First-Year Engineering Design Education. In Tucker, R. (Ed.), *Collaboration and Student Engagement in Design Education* (pp. 145–168). Hershey, PA: IGI Global. DOI: 10.4018/978-1-5225-0726-0.ch007

Panchenko, V. (2021). Prospects for Energy Supply of the Arctic Zone Objects of Russia Using Frost-Resistant Solar Modules. In P. Vasant, G. Weber, & W. Punurai (Eds.), *Research Advancements in Smart Technology, Optimization, and Renewable Energy* (pp. 149-169). IGI Global. https://doi.org/10.4018/978-1-7998-3970-5.ch008

Panchenko, V. (2021). Photovoltaic Thermal Module With Paraboloid Type Solar Concentrators. *International Journal of Energy Optimization and Engineering*, *10*(2), 1–23. https://doi.org/10.4018/IJEOE.2021040101

Pandey, K., & Datta, S. (2021). Dry Machining of Inconel 825 Superalloys: Performance of Tool Inserts (Carbide, Cermet, and SiAlON). *International Journal of Manufacturing, Materials, and Mechanical Engineering*, *11*(4), 26–39. DOI: 10.4018/IJMMME.2021100102

Panneer, R. (2017). Effect of Composition of Fibers on Properties of Hybrid Composites. *International Journal of Manufacturing, Materials, and Mechanical Engineering*, *7*(4), 28–43. DOI: 10.4018/IJMMME.2017100103

Pany, C. (2021). Estimation of Correct Long-Seam Mismatch Using FEA to Compare the Measured Strain in a Non-Destructive Testing of a Pressurant Tank: A Reverse Problem. *International Journal of Smart Vehicles and Smart Transportation*, *4*(1), 16–28. DOI: 10.4018/IJSVST.2021010102

Paul, S., & Roy, P. (2018). Optimal Design of Power System Stabilizer Using a Novel Evolutionary Algorithm. *International Journal of Energy Optimization and Engineering*, *7*(3), 24–46. DOI: 10.4018/IJEOE.2018070102

Paul, S., & Roy, P. K. (2021). Oppositional Differential Search Algorithm for the Optimal Tuning of Both Single Input and Dual Input Power System Stabilizer. In P. Vasant, G. Weber, & W. Punurai (Eds.), *Research Advancements in Smart Technology, Optimization, and Renewable Energy* (pp. 256-282). IGI Global. https://doi.org/10.4018/978-1-7998-3970-5.ch013

Pavaloiu, A. (2018). Artificial Intelligence Ethics in Biomedical-Engineering-Oriented Problems. In Kose, U., Guraksin, G., & Deperlioglu, O. (Eds.), *Nature-Inspired Intelligent Techniques for Solving Biomedical Engineering Problems* (pp. 219–231). Hershey, PA: IGI Global. DOI: 10.4018/978-1-5225-4769-3.ch010

Pioro, I., Mahdi, M., & Popov, R. (2017). Application of Supercritical Pressures in Power Engineering. In Chen, L., & Iwamoto, Y. (Eds.), *Advanced Applications of Supercritical Fluids in Energy Systems* (pp. 404–457). Hershey, PA: IGI Global. DOI: 10.4018/978-1-5225-2047-4.ch013

Plaksina, T., & Gildin, E. (2017). Rigorous Integrated Evolutionary Workflow for Optimal Exploitation of Unconventional Gas Assets. *International Journal of Energy Optimization and Engineering*, *6*(1), 101–122. DOI: 10.4018/IJEOE.2017010106

Popat, J., Kakadiya, H., Tak, L., Singh, N. K., Majeed, M. A., & Mahajan, V. (2021). Reliability of Smart Grid Including Cyber Impact: A Case Study. In Singh, R., Singh, A., Dwivedi, A., & Nagabhushan, P. (Eds.), *Computational Methodologies for Electrical and Electronics Engineers* (pp. 163–174). IGI Global. https://doi.org/10.4018/978-1-7998-3327-7.ch013

Quiza, R., La Fé-Perdomo, I., Rivas, M., & Ramtahalsing, V. (2021). Triple Bottom Line-Focused Optimization of Oblique Turning Processes Based on Hybrid Modeling: A Study Case on AISI 1045 Steel Turning. In Burstein, L. (Ed.), *Handbook of Research on Advancements in Manufacturing, Materials, and Mechanical Engineering* (pp. 215–241). IGI Global. https://doi.org/10.4018/978-1-7998-4939-1.ch010

Rahmani, M. K. (2022). Blockchain Technology: Principles and Algorithms. In Khan, S., Syed, M., Hammad, R., & Bushager, A. (Eds.), *Blockchain Technology and Computational Excellence for Society 5.0* (pp. 16–27). IGI Global. https://doi.org/10.4018/978-1-7998-8382-1.ch002

Ramdani, N., & Azibi, M. (2018). Polymer Composite Materials for Microelectronics Packaging Applications: Composites for Microelectronics Packaging. In Kumar, K., & Davim, J. (Eds.), *Composites and Advanced Materials for Industrial Applications* (pp. 177–211). Hershey, PA: IGI Global. DOI: 10.4018/978-1-5225-5216-1.ch009

Ramesh, M., Garg, R., & Subrahmanyam, G. V. (2017). Investigation of Influence of Quenching and Annealing on the Plane Fracture Toughness and Brittle to Ductile Transition Temperature of the Zinc Coated Structural Steel Materials. *International Journal of Surface Engineering and Interdisciplinary Materials Science*, *5*(2), 33–42. DOI: 10.4018/IJSEIMS.2017070103

Robinson, J., & Beneroso, D. (2022). Project-Based Learning in Chemical Engineering: Curriculum and Assessment, Culture and Learning Spaces. In Alves, A., & van Hattum-Janssen, N. (Eds.), *Training Engineering Students for Modern Technological Advancement* (pp. 1–19). IGI Global. https://doi.org/10.4018/978-1-7998-8816-1.ch001

Rondon, B. (2021). Experimental Characterization of Admittance Meter With Crude Oil Emulsions. *International Journal of Electronics, Communications, and Measurement Engineering*, *10*(2), 51–59. https://doi.org/10.4018/IJECME.2021070104

Rudolf, S., Biryuk, V. V., & Volov, V. (2018). Vortex Effect, Vortex Power: Technology of Vortex Power Engineering. In Kharchenko, V., & Vasant, P. (Eds.), *Handbook of Research on Renewable Energy and Electric Resources for Sustainable Rural Development* (pp. 500–533). Hershey, PA: IGI Global. DOI: 10.4018/978-1-5225-3867-7.ch021

Sah, A., Bhadula, S. J., Dumka, A., & Rawat, S. (2018). A Software Engineering Perspective for Development of Enterprise Applications. In Elçi, A. (Ed.), *Handbook of Research on Contemporary Perspectives on Web-Based Systems* (pp. 1–23). Hershey, PA: IGI Global. DOI: 10.4018/978-1-5225-5384-7.ch001

Sahli, Y., Zitouni, B., & Hocine, B. M. (2021). Three-Dimensional Numerical Study of Overheating of Two Intermediate Temperature P-AS-SOFC Geometrical Configurations. In Badea, G., Felseghi, R., & A chilean, I. (Eds.), *Hydrogen Fuel Cell Technology for Stationary Applications* (pp. 186–222). IGI Global. https://doi.org/10.4018/978-1-7998-4945-2.ch008

Sahoo, P., & Roy, S. (2017). Tribological Behavior of Electroless Ni-P, Ni-P-W and Ni-P-Cu Coatings: A Comparison. *International Journal of Surface Engineering and Interdisciplinary Materials Science*, *5*(1), 1–15. DOI: 10.4018/IJSEIMS.2017010101

Sahoo, S. (2018). Laminated Composite Hypar Shells as Roofing Units: Static and Dynamic Behavior. In Kumar, K., & Davim, J. (Eds.), *Composites and Advanced Materials for Industrial Applications* (pp. 249–269). Hershey, PA: IGI Global. DOI: 10.4018/978-1-5225-5216-1.ch011

Sahu, H., & Hungyo, M. (2018). Introduction to SDN and NFV. In Dumka, A. (Ed.), *Innovations in Software-Defined Networking and Network Functions Virtualization* (pp. 1–25). Hershey, PA: IGI Global. DOI: 10.4018/978-1-5225-3640-6.ch001

Salem, A. M., & Shmelova, T. (2018). Intelligent Expert Decision Support Systems: Methodologies, Applications, and Challenges. In Shmelova, T., Sikirda, Y., Rizun, N., Salem, A., & Kovalyov, Y. (Eds.), *Socio-Technical Decision Support in Air Navigation Systems: Emerging Research and Opportunities* (pp. 215–242). Hershey, PA: IGI Global. DOI: 10.4018/978-1-5225-3108-1.ch007

Samal, M. (2017). FE Analysis and Experimental Investigation of Cracked and Un-Cracked Thin-Walled Tubular Components to Evaluate Mechanical and Fracture Properties. In Samui, P., Chakraborty, S., & Kim, D. (Eds.), *Modeling and Simulation Techniques in Structural Engineering* (pp. 266–293). Hershey, PA: IGI Global. DOI: 10.4018/978-1-5225-0588-4.ch009

Samal, M., & Balakrishnan, K. (2017). Experiments on a Ring Tension Setup and FE Analysis to Evaluate Transverse Mechanical Properties of Tubular Components. In Samui, P., Chakraborty, S., & Kim, D. (Eds.), *Modeling and Simulation Techniques in Structural Engineering* (pp. 91–115). Hershey, PA: IGI Global. DOI: 10.4018/978-1-5225-0588-4.ch004

Samarasinghe, D. A., & Wood, E. (2021). Innovative Digital Technologies. In Underwood, J., & Shelbourn, M. (Eds.), *Handbook of Research on Driving Transformational Change in the Digital Built Environment* (pp. 142–163). IGI Global. https://doi.org/10.4018/978-1-7998-6600-8.ch006

Sawant, S. (2018). Deep Learning and Biomedical Engineering. In Kose, U., Guraksin, G., & Deperlioglu, O. (Eds.), *Nature-Inspired Intelligent Techniques for Solving Biomedical Engineering Problems* (pp. 283–296). Hershey, PA: IGI Global. DOI: 10.4018/978-1-5225-4769-3.ch014

Schulenberg, T. (2021). Energy Conversion Using the Supercritical Steam Cycle. In Chen, L. (Ed.), *Handbook of Research on Advancements in Supercritical Fluids Applications for Sustainable Energy Systems* (pp. 659–681). IGI Global. DOI: 10.4018/978-1-7998-5796-9.ch018

Sezgin, H., & Berkalp, O. B. (2018). Textile-Reinforced Composites for the Automotive Industry. In Kumar, K., & Davim, J. (Eds.), *Composites and Advanced Materials for Industrial Applications* (pp. 129–156). Hershey, PA: IGI Global. DOI: 10.4018/978-1-5225-5216-1.ch007

Shaaban, A. A., & Shehata, O. M. (2021). Combining Response Surface Method and Metaheuristic Algorithms for Optimizing SPIF Process. *International Journal of Manufacturing, Materials, and Mechanical Engineering, 11*(4), 1–25. https://doi.org/10.4018/IJMMME.2021100101

Shafaati Shemami, M., & Sefid, M. (2022). Implementation and Demonstration of Electric Vehicle-to-Home (V2H) Application: A Case Study. In Alam, M., Pillai, R., & Murugesan, N. (Eds.), *Developing Charging Infrastructure and Technologies for Electric Vehicles* (pp. 268–293). IGI Global. https://doi.org/10.4018/978-1-7998-6858-3.ch015

Shah, M. Z., Gazder, U., Bhatti, M. S., & Hussain, M. (2018). Comparative Performance Evaluation of Effects of Modifier in Asphaltic Concrete Mix. *International Journal of Strategic Engineering*, *1*(2), 13–25. DOI: 10.4018/IJoSE.2018070102

Sharma, N., & Kumar, K. (2018). Fabrication of Porous NiTi Alloy Using Organic Binders. In Kumar, K., & Davim, J. (Eds.), *Composites and Advanced Materials for Industrial Applications* (pp. 38–62). Hershey, PA: IGI Global. DOI: 10.4018/978-1-5225-5216-1.ch003

Shivach, P., Nautiyal, L., & Ram, M. (2018). Applying Multi-Objective Optimization Algorithms to Mechanical Engineering. In Ram, M., & Davim, J. (Eds.), *Soft Computing Techniques and Applications in Mechanical Engineering* (pp. 287–301). Hershey, PA: IGI Global. DOI: 10.4018/978-1-5225-3035-0.ch014

Shmelova, T. (2018). Stochastic Methods for Estimation and Problem Solving in Engineering: Stochastic Methods of Decision Making in Aviation. In Kadry, S. (Ed.), *Stochastic Methods for Estimation and Problem Solving in Engineering* (pp. 139–160). Hershey, PA: IGI Global. DOI: 10.4018/978-1-5225-5045-7.ch006

Siero González, L. R., & Romo Vázquez, A. (2017). Didactic Sequences Teaching Mathematics for Engineers With Focus on Differential Equations. In Ramírez-Montoya, M. (Ed.), *Handbook of Research on Driving STEM Learning With Educational Technologies* (pp. 129–151). Hershey, PA: IGI Global. DOI: 10.4018/978-1-5225-2026-9.ch007

Sim, M. S., You, K. Y., Esa, F., & Chan, Y. L. (2021). Nanostructured Electromagnetic Metamaterials for Sensing Applications. In Bhat, M., Wani, I., & Ashraf, S. (Eds.), *Applications of Nanomaterials in Agriculture, Food Science, and Medicine* (pp. 141–164). IGI Global. https://doi.org/10.4018/978-1-7998-5563-7.ch009

Singh, R., & Dutta, S. (2018). Visible Light Active Nanocomposites for Photocatalytic Applications. In Kumar, K., & Davim, J. (Eds.), *Composites and Advanced Materials for Industrial Applications* (pp. 270–296). Hershey, PA: IGI Global. DOI: 10.4018/978-1-5225-5216-1.ch012

Skripov, P. V., Yampol'skiy, A. D., & Rutin, S. B. (2021). High-Power Heat Transfer in Supercritical Fluids: Microscale Times and Sizes. In Chen, L. (Ed.), *Handbook of Research on Advancements in Supercritical Fluids Applications for Sustainable Energy Systems* (pp. 424–450). IGI Global. https://doi.org/10.4018/978-1-7998-5796-9.ch012

Sözbilir, H., Özkaymak, Ç., Uzel, B., & Sümer, Ö. (2018). Criteria for Surface Rupture Microzonation of Active Faults for Earthquake Hazards in Urban Areas. In Ceryan, N. (Ed.), *Handbook of Research on Trends and Digital Advances in Engineering Geology* (pp. 187–230). Hershey, PA: IGI Global. DOI: 10.4018/978-1-5225-2709-1.ch005

Stanciu, I. (2018). Stochastic Methods in Microsystems Engineering. In Kadry, S. (Ed.), *Stochastic Methods for Estimation and Problem Solving in Engineering* (pp. 161–176). Hershey, PA: IGI Global. DOI: 10.4018/978-1-5225-5045-7.ch007

Strebkov, D., Nekrasov, A., Trubnikov, V., & Nekrasov, A. (2018). Single-Wire Resonant Electric Power Systems for Renewable-Based Electric Grid. In Kharchenko, V., & Vasant, P. (Eds.), *Handbook of Research on Renewable Energy and Electric Resources for Sustainable Rural Development* (pp. 449–474). Hershey, PA: IGI Global. DOI: 10.4018/978-1-5225-3867-7.ch019

Sukhyy, K., Belyanovskaya, E., & Sukhyy, M. (2021). *Basic Principles for Substantiation of Working Pair Choice*. IGI Global. DOI: 10.4018/978-1-7998-4432-7.ch002

Suri, M. S., & Kaliyaperumal, D. (2022). Extension of Aspiration Level Model for Optimal Planning of Fast Charging Stations. In Fekik, A., & Benamrouche, N. (Eds.), *Modeling and Control of Static Converters for Hybrid Storage Systems* (pp. 91–106). IGI Global. https://doi.org/10.4018/978-1-7998-7447-8.ch004

Tallet, E., Gledson, B., Rogage, K., Thompson, A., & Wiggett, D. (2021). Digitally-Enabled Design Management. In Underwood, J., & Shelbourn, M. (Eds.), *Handbook of Research on Driving Transformational Change in the Digital Built Environment* (pp. 63–89). IGI Global. https://doi.org/10.4018/978-1-7998-6600-8.ch003

Terki, A., & Boubertakh, H. (2021). A New Hybrid Binary-Real Coded Cuckoo Search and Tabu Search Algorithm for Solving the Unit-Commitment Problem. *International Journal of Energy Optimization and Engineering, 10*(2), 104–119. https://doi.org/10.4018/IJEOE.2021040105

Tüdeş, Ş., Kumlu, K. B., & Ceryan, S. (2018). Integration Between Urban Planning and Natural Hazards For Resilient City. In Ceryan, N. (Ed.), *Handbook of Research on Trends and Digital Advances in Engineering Geology* (pp. 591–630). Hershey, PA: IGI Global. DOI: 10.4018/978-1-5225-2709-1.ch017

Ulamis, K. (2018). Soil Liquefaction Assessment by Anisotropic Cyclic Triaxial Test. In Ceryan, N. (Ed.), *Handbook of Research on Trends and Digital Advances in Engineering Geology* (pp. 631–664). Hershey, PA: IGI Global. DOI: 10.4018/978-1-5225-2709-1.ch018

Valente, M., & Milani, G. (2017). Seismic Assessment and Retrofitting of an Under-Designed RC Frame Through a Displacement-Based Approach. In Plevris, V., Kremmyda, G., & Fahjan, Y. (Eds.), *Performance-Based Seismic Design of Concrete Structures and Infrastructures* (pp. 36–58). Hershey, PA: IGI Global. DOI: 10.4018/978-1-5225-2089-4.ch002

Vargas-Bernal, R. (2021). Advances in Electromagnetic Environmental Shielding for Aeronautics and Space Applications. In Nikolopoulos, C. (Ed.), *Recent Trends on Electromagnetic Environmental Effects for Aeronautics and Space Applications* (pp. 80–96). IGI Global. https://doi.org/10.4018/978-1-7998-4879-0.ch003

Vasant, P. (2018). A General Medical Diagnosis System Formed by Artificial Neural Networks and Swarm Intelligence Techniques. In Kose, U., Guraksin, G., & Deperlioglu, O. (Eds.), *Nature-Inspired Intelligent Techniques for Solving Biomedical Engineering Problems* (pp. 130–145). Hershey, PA: IGI Global. DOI: 10.4018/978-1-5225-4769-3.ch006

Verner, C. M., & Sarwar, D. (2021). Avoiding Project Failure and Achieving Project Success in NHS IT System Projects in the United Kingdom. *International Journal of Strategic Engineering*, *4*(1), 33–54. https://doi.org/10.4018/IJoSE.2021010103

Verrollot, J., Tolonen, A., Harkonen, J., & Haapasalo, H. J. (2018). Challenges and Enablers for Rapid Product Development. *International Journal of Applied Industrial Engineering*, *5*(1), 25–49. DOI: 10.4018/IJAIE.2018010102

Wan, A. C., Zulu, S. L., & Khosrow-Shahi, F. (2021). Industry Views on BIM for Site Safety in Hong Kong. In Underwood, J., & Shelbourn, M. (Eds.), *Handbook of Research on Driving Transformational Change in the Digital Built Environment* (pp. 120–140). IGI Global. https://doi.org/10.4018/978-1-7998-6600-8.ch005

Yardimci, A. G., & Karpuz, C. (2018). Fuzzy Rock Mass Rating: Soft-Computing-Aided Preliminary Stability Analysis of Weak Rock Slopes. In Ceryan, N. (Ed.), *Handbook of Research on Trends and Digital Advances in Engineering Geology* (pp. 97–131). Hershey, PA: IGI Global. DOI: 10.4018/978-1-5225-2709-1.ch003

You, K. Y. (2021). Development Electronic Design Automation for RF/Microwave Antenna Using MATLAB GUI. In Nwajana, A., & Ihianle, I. (Eds.), *Handbook of Research on 5G Networks and Advancements in Computing, Electronics, and Electrical Engineering* (pp. 70–148). IGI Global. https://doi.org/10.4018/978-1-7998-6992-4.ch004

Yousefi, Y., Gratton, P., & Sarwar, D. (2021). Investigating the Opportunities to Improve the Thermal Performance of a Case Study Building in London. *International Journal of Strategic Engineering*, *4*(1), 1–18. https://doi.org/10.4018/IJoSE.2021010101

Zindani, D., & Kumar, K. (2018). Industrial Applications of Polymer Composite Materials. In Kumar, K., & Davim, J. (Eds.), *Composites and Advanced Materials for Industrial Applications* (pp. 1–15). Hershey, PA: IGI Global. DOI: 10.4018/978-1-5225-5216-1.ch001

Zindani, D., Maity, S. R., & Bhowmik, S. (2018). A Decision-Making Approach for Material Selection of Polymeric Composite Bumper Beam. In Kumar, K., & Davim, J. (Eds.), *Composites and Advanced Materials for Industrial Applications* (pp. 112–128). Hershey, PA: IGI Global. DOI: 10.4018/978-1-5225-5216-1.ch006

# About the Contributors

**Tolga Taner** is a Professor at Aksaray University in Türkiye. He has been the Head of Motor Vehicles and Transportation Technologies since 2012. He has been working as an academic and industrial engineering for 26 years. He has many academic articles (SCI index) and books in the fields of energy, exergy, biogas, renewable energy, hydrogen, and fuel cell. He is an Editor and a reviewer for many journals (SCI index). He works as a Mentor/Supervisor in some industrial factories. After graduating from the Department of Mechanical Engineering in 1998, he received his PhD in Mechanical Engineering from the Faculty of Engineering of Ankara Gazi University in 2013. He worked as a Lecturer (Part-Time) in the Middle East Technical (METU) between 2003-2006. In 2020, 2021, 2022, 2023, and 2024 (5th times), he was listed in the "2% World's Most Influential Scientists", "Mechanical Engineering-Energy field" field published by Stanford University. He has taken part in many research projects as a coordinator and researcher.

**Hafiz Muhammad Ali** presently serves as an associate professor and director of the renewable energy laboratory at the Mechanical Engineering Department at King Fahd University of Petroleum and Minerals, Saudi Arabia. He possesses a PhD degree in mechanical engineering from the School of Engineering and Materials Science at Queen Mary University of London, UK. Dr. Ali served as a postdoctoral researcher at the Water and Energy Laboratory at the University of California, Merced, United States. Dr. Ali conducted substantial research in thermal engineering and served as the director of the Thermal and Clean Energy Research

Laboratory, the Renewable Energy Research and Development Centre, and the Heat & Mass Transfer and Refrigeration Laboratory at the University of Engineering and Technology Taxila, Pakistan. He has authored over 300 articles, numerous books, and book chapters. He has actively engaged in delivering keynote and invited presentations at numerous international conferences. Dr. Ali has received numerous worldwide accolades, particularly the Highly Cited Researcher Award from Clarivate. He serves as an editorial board member for multiple international journals, including the Journal of Thermal Analysis and Calorimetry, Applied Nanoscience, Energy Resources: Part A, Heat Transfer, Journal of Thermal Science, and the Journal of Mechanical Engineering.

***

**Dharmbir Prasad** was born in 1986 at Nalanda, Bihar, India. He received his B.Tech. degree in electrical engineering from Hooghly Engineering & Technology College (under West Bengal University of Technology), Hooghly, India and M.Tech. in Power System from Dr. B.C. Roy Engineering College (under West Bengal University of Technology), Durgapur, India, respectively. Currently, he is working in the capacity of assistant professor in the department of electrical engineering, Asansol Engineering College, Asansol, West Bengal, India. Now, he is pursuing his Ph.D. degree from Indian School of Mines, Dhanbad, Jharkhand, India. His research interest includes economic operation of power system.

**Ranadip Roy**, seeking opportunities for further research and academic collaboration Experienced Ph.D. holder from Indian Institute of Technology with over 12 years in academia, specializing in renewable energy optimization, soft computing, and power system optimization. Currently an Associate Professor and Dean of Students' Affairs at Sanaka Educational Trust's Group of Institutions, Durgapur, West Bengal. Recognized for receiving the Best Researcher Award twice and as the Best Coordinator, Eastern Region (MoE, NDLI, IIT Kharagpur). Actively involved in Ministry of Education initiatives as an Evaluator (SIH) and Innovation Ambassador for MoE, India. More than 30+ research papers, book chapters, patent and conferences at various SCI, Scopus and IEEE indexed journals and other platforms. Skilled reviewer for esteemed journals and IEEE conferences.

**Rudra Pratap Singh** was born in 1983 at Chittaranjan, Burdwan, West Bengal, India. He received his B. Tech. and M. Tech degree in electrical engineering from Asansol Engineering (under the West Bengal University of Technology), Asansol, West Bengal, India and MBA in Power Management from the University of Petroleum and Energy Studies, Dehradun, Uttrakhand, India, respectively. He received his Ph.D. degree from the Indian School of Mines, Dhanbad, Jharkhand, India. Currently, he is working in the capacity of assistant professor in the department of Electrical Engineering, Asansol Engineering College, Asansol, West Bengal, India. His research interest includes the application of state estimation and optimization techniques in various fields of engineering.

**Rahul Uday Urunkar** is a committed and result-oriented academician specializing in Mechanical Design and Thermal Engineering. With a strong background in teaching, research, and academic administration, he has contributed as an editor of a technical book and as a reviewer for several international journals. Dr. Urunkar has delivered expert sessions in various Faculty Development Programs and organized a notable Student Development Program on UPSC career opportunities. He possesses proficiency in multiple proprietary software tools and has actively participated in numerous FDPs, STTPs, and workshops sponsored by AICTE, ISTE, ISHRAE, TEQIP, IEEE, and IE Kolkata.

# Index

www.ingramcontent.com/pod-product-compliance
Lightning Source LLC
LaVergne TN
LVHW081326300926
849161LV00008B/1892